Hayfaa Jasim Hussein
Mohamed Abdel-Raheem

Mineral and Organic Fertilization and Its Effects on the Environment

Hayfaa Jasim Hussein
Mohamed Abdel-Raheem

Mineral and Organic Fertilization and Its Effects on the Environment

Noor Publishing

Imprint

Any brand names and product names mentioned in this book are subject to trademark, brand or patent protection and are trademarks or registered trademarks of their respective holders. The use of brand names, product names, common names, trade names, product descriptions etc. even without a particular marking in this work is in no way to be construed to mean that such names may be regarded as unrestricted in respect of trademark and brand protection legislation and could thus be used by anyone.

Cover image: www.ingimage.com

Publisher:
Noor Publishing
is a trademark of
Dodo Books Indian Ocean Ltd. and OmniScriptum S.R.L publishing group

120 High Road, East Finchley, London, N2 9ED, United Kingdom
Str. Armeneasca 28/1, office 1, Chisinau MD-2012, Republic of Moldova, Europe
Managing Directors: Ieva Konstantinova, Victoria Ursu
info@omniscriptum.com

Printed at: see last page
ISBN: 978-620-5-63713-5

Mineral and Organic Fertilization and Its Effects on the Environment

By

Prof. Dr. Hayfaa Jasim Hussein

Prof. Dr. Mohamed Abdel-Raheem

(2023)

Prof. Dr. Hayfaa Jasim Hussein

Department of Soil Science and Water Resources, College of Agriculture, University of Basrah, Iraq

E-mail: hayfaa.hussein@uobasrah.edu.iq

Tell: (+964) 7703162753

Prof. Dr. Mohamed Abdel-Raheem Ali Abdel-Raheem

Professor of Entomology (Biological Control)

Pests & Plant Protection Department

Agricultural and Biological Research Institute

National Research Centre.

33rd ElBohouth St., Dokki, Cairo, Egypt.

Email: abdelraheem_nrc@hotmail.com,

abdelraheem_nrc@yahoo.com

Mobile: (+2) 01155527583 - (+2) 01009580797

Content list

Introduction

The intensive use of mineral (M) fertilizers may cause harm the environment via leaching or greenhouse gas emissions, destroy soil fertility as a consequence of loss of soil organic matter, and, due to their high price, they are economically unviable for producers. It is widely accepted that organic (O) fertilizers may deal with pressing challenges facing modern agriculture, even if farmers need to improve their knowledge for applying in fertilization programs.

A meta-analysis approach has been adopted to evaluate the effects on soil organic carbon (SOC) and crop yield of O fertilizers, applied alone or in combination with mineral fertilizers (MO) under conventional (CT), reduced (RT), and no-tillage (NT) regimes. The analysis was performed in different climatic conditions, soil properties, crop species, and irrigation management. Organic fertilizers have a positive influence in increasing SOC compared with M (on average 12.9%), even if high values were observed under NT (20.6%).

The need for flexible and environment-specific systems when considering organic fertilization subjected to different tillage regimes. Similarly, MO application showed a better crop yield response in CT and RT under coarse soils when compared with M fertilizer applied alone (on average 13.4 and 12.7%, respectively), while in medium-textured soils, CT and RT yielded better than NT under O fertilizers (9.5 and 11.2 vs. 2.5%, respectively).

Among the crop species, legumes performed better when O fertilizers were adopted than M fertilizers (on average 15.2%), while among the other crop species, few differences were detected among the fertilization programs. Under irrigated systems, RT and NT led to higher productivity than CT, especially under MO treatments (on average 9.2 vs. 3.4%, respectively).

Soil microorganisms are one of the main indicators used for assessing the stability of the soil ecosystem, the metabolism in the soil, and its fertility. The most important are the active soil microorganisms and the influence of the fertilizer applied to the soil on the abundance of these microorganisms.

We aimed to investigate how the applied organic fertilizers affect the most active soil microorganisms, which determine the soil fertility and stability. Fungi, yeast-like fungi abundance, and abundance of three physiological groups of bacteria were analyzed: non-symbiotic diazotrophic, organotrophic, and mineral nitrogen assimilating.

The effect of fertilizers on trends in the changes of microorganism community diversity; however, more analysis is needed to assess the impact of organic fertilizers on the most active soil microorganisms. Therefore, the investigation was continued. The results of the 2020 quantitative analysis of culturable soil microorganisms show that the highest abundance of organotrophic and non-symbiotic diazotrophic bacteria were recorded during the summer season. Meanwhile, the abundance of bacteria assimilating mineral nitrogen and fungi was higher in autumn. Agrochemical parameters were determined at the beginning of the experiment.

The highest concentration of N_{min} in the soil was determined after fertilizing the plants with the combination of granulated poultry manure (N_{170}) + biological substance *Azotobacter* spp.

The yield of barley was calculated. It was found that the highest yield of spring barley in 2020 was obtained by fertilizing the experimental field with organic in combination with mineral fertilizers.

Incredible achievements have been made in agricultural production worldwide, but many daunting challenges remain unresolved to ensure food security and environmental sustainability.

Chemical fertilizers are used in excessive and disproportionate quantities to raise crop yields in order to combat certain circumstances. However, apart from being processed in crop plants, chemical fertilizers above the threshold level pollute the atmosphere.

As the availability of nutrients becomes a constraint of plant growth and production, sustained crop productivity relies on constant renewal. To increase agriculture production and maintain soil fertility, the application of chemical fertilizers is indispensable.

However, insufficient or unnecessary application of fertilizer does not guarantee consistently growing yields, which can result in low efficiency of nutrient usage. Today, the key goals are the study of the effective use of chemicals, the reduction of production costs and the efficient use of fertilization.

In modern agriculture, besides providing high and stable yields, it is imperative to produce products with a high nutritive quality. The goal of this study was to determine the effect of different fertilization regimes on the macro- and micronutrients in beetroot.

A 3-year field trial was set up according to a Latin square method with four types of fertilization (unfertilized control, 50 t stable manure ha^{-1}, and 500 and 1,000 kg NPK 5-20-30 ha^{-1}). The mineral content was determined as follows (mg 100 g^{-1} in fresh weight of beetroot): 14–29 P, 189–354 K, 18–34 Ca, 17–44 Mg, 0.67–1.83 Fe, 0.41–0.65 Mn and 0.28–0.44 Zn.

The highest beetroot P content was determined for the treatment with stable manure, especially in a year with dry climatic conditions. The highest beetroot K content was determined for the treatment with 1,000 kg NPK 5-20-30 ha^{-1}, but at the same time for the same treatment, a general decreasing trend of micronutrient content was determined, due to the possible antagonistic effect of added potassium.

For better mineral status of beetroot, application of combined mineral and organic fertilizers supplemented with additional foliar application of micronutrients can be suggested.

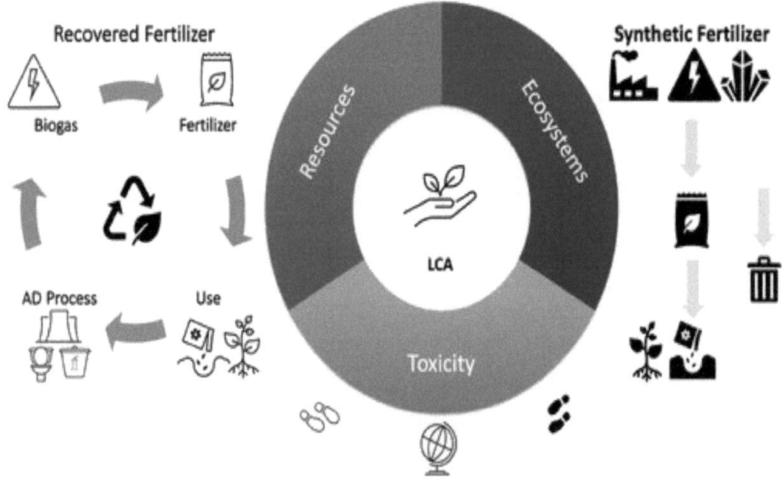

Recovered fertilizers (RFs), in the form of digestate and digestate-derived ammonium sulfate, were produced from organic wastes by thermophilic anaerobic digestion (AD) at full scale. RFs were then used for crop production (maize), substituting synthetic mineral fertilizers (SFs).

Environmental impacts due to both RF and SF production and use were studied by a life cycle assessment (LCA) approach using, as much as possible, data directly measured at full scale. The functional unit chosen was referred to as the fertilization of 1 ha of maize, as this paper intends to investigate the impacts of the use of RF (Scenario RF) for crop fertilization compared to that of SF (Scenario SF).

Scenario RF showed better environmental performances than the system encompassing the production and use of urea and synthetic fertilizers (Scenario SF). In particular, for the Scenario RF, 11 of the 18 categories showed a lower impact than the Scenario SF, and 3 of the categories (ionizing radiation, fossil resource scarcity, and water consumption) showed net negative impacts in Scenario RF, getting the benefits from the credit for renewable energy production by AD.

The LCA approach also allowed proposing precautions able to reduce further fertilizer impacts, resulting in total negative impacts in using RF for crop production. Anaerobic digestion represents the key to propose a sustainable approach in producing renewable fertilizers, thanks to both energy production and the modification that occurs to waste during a biological process, leaving a substrate (digestate) with high amending and fertilizing properties.

The growing population associated with rising hunger stimulated the industrialization of agricultural practices that require increased use of farmland to produce the highest yield by means of intensive use mineral fertilizers and heavy soil tillage. Adopting these practices led to significant growth in agricultural productivity but came at the cost of environmental and soil health [1,2].

Indeed, the intensive applications of mineral fertilizer caused loss of soil organic carbon, environmental pollution, overexploitation of natural resources, loss of biodiversity, and adverse climate changes. In addition, chemical fertilizers are also criticized due to their effect on soil deterioration as their intensive use in agroecosystems to improve crop productivity causes a gradual modification of soil physical properties making soils acidic [3].

Recently, the concept of sustainable agriculture has been developed as a system of ecological farming practices based on scientific innovations to satisfy the needs of humankind for healthy foods, while maintaining the quality of the environment and the natural resources base. The adoption of organic fertilizers instead of mineral ones could represent an environment-friendly practice through sustainable farming systems. Indeed, organic fertilizers are obtained from organic materials (i.e., plant residues, animal manures, by-products of the food industry) and subjected to several processes, such as fermentation, drying, chopping, and composting [4,5].

It has been reported that using organic nutrient sources associated with natural pest control in diverse cropping systems will lead to better agroecosystem services [6]. In addition, plant residues or animal waste used have been used as organic fertilizers due to their richness of organic matter, improving soil structure and helping microbes thrive, and consequently providing nutrients to crop plants in a natural biological process. However, one of the main difficulties related to organic fertilizers is that since organic materials are variable in terms of quantity and quality, their content in macro and micro-nutrients are lower compared to inorganic fertilizers. This requires the adoption of excessive amounts of organic materials to respond to the nutrient demand to support high crop yields [7].

The reduction of soil carbon content observed in the agricultural fields has been due also to intensive and frequent tillage practices, which contribute to creating an aerobic environment in the soil that facilitates the mineralization rate of soil organic matter [8]. Although soil tillage is used for seedbed preparation, weed suppression, soil aeration, the management of crop residue and cover crop biomass, leveling the soil, incorporating manure and fertilizer into the root zone, intensive tillage passes fractures the soil, it disrupts the soil structure, facilitating soil crusting and erosion [9,10].

Conservation tillage represents another approach to sustainable agriculture that has been continuously evolving since the late 20th century with the main goal to address problems, such as soil erosion and degradation, meanwhile preserving natural resources. Nowadays, conservation tillage is one of the main pillars of conservation agriculture farming aimed to protect soils by permanent soil cover through previous crop residues or cover crops and reduce soil tillage process in diverse cropping systems.

Several studies showed that the application of conservation tillage would lead to improving soil health by increasing water and nutrient use efficiency and reducing soil degradation. Moreover, agroecosystems managed under conservation agriculture practices may prevent further losses; at the same time, conservation can be used to re-establish degraded lands. Conservation agriculture has been practiced and promoted worldwide by applying conservation tillage practices, particularly because of expected benefits such as a decreased cost, labour intensity, and reduced greenhouse gases emissions. In addition, the FAO has reported that conservation agriculture is practiced worldwide, with an increase in the total area from 45 million ha in 1999 to over 200 million hectares in 2021 [11].

It is interesting to note that about half the areas adopting conservation agriculture are located in developing countries [12]. The reduction of tillage depth through reduced-tillage (RT) or no-tillage (NT) practices represents sustainable strategies of conservation agriculture. Overall conservation tillage systems protect soils, conserve soil moisture, and get the best benefits of previous crop residues.

Integrating organic fertilizer application with conservation tillage practices could be a suitable option for sustainable agriculture.

Various studies have been conducted to evaluate how sustainable farming management could be adopted to replace conventional or modern industrial farming and try to understand how different management systems may influence environmental health, soil structure and fertility, and crop productivity. Several authors report a reduction in greenhouse gas emissions [13,14], an improvement of soil quality [15,16], and, consequently, an enhancement of soil fertility [7,17], all while better maintaining biodiversity [18,19].

Despite that, often yield gap is reported as the main challenge for adapting these sustainable farming managements. Research reports confirmed that under certain conditions, the yields of sustainable and conventional farming practices could be equal, even if the effects of organic fertilization and conservation tillage practices are not well understood and widely accepted by farmers, given the wide range of conditions the agricultural activities are subjected.

Improving knowledge concerning the combination of organic fertilizers with tillage regimes may represent an important decision tool for the farmers that will arrange their activities based on the interaction of these farming practices with climatic conditions, crop rotation, crop choice, and soil characteristics. This meta-analysis study has been conducted to identify the implication of crop fertilization and soil tillage on crop yield and soil organic carbon (SOC). The study hypothesizes that farmer decisions, in terms of fertilization source and soil tillage, are variable and should be well integrated with the agricultural and environmental factors that characterize the agroecosystems. For this purpose, this study compared the use of organic fertilizers and conservation tillage (minimum tillage/no-tillage) related to mineral fertilizers and conventional tillage practices, respectively.

Soil characteristics are subjected to important modifications based on the tillage system adopted; therefore, the choice of appropriate fertilizer source and tillage operations should be addressed based on climate conditions, soil conditions, type of crops, and management factors [20]. The results of subgroup analysis for each categorical variable are separately presented in this study.

The differences of using organic fertilizer (O) or the combination of mineral plus organic fertilizers (MO) instead of mineral fertilizers (M) under different tillage systems on soil organic carbon (SOC) are reported in **Figure 1**. The soil organic carbon represents a key factor for the soil fertility of agroecosystems, and all agricultural practices that affect the modification of SOC in the agricultural soil should be carefully considered to avoid dangerous environmental changes [21]. Improving the content of organic matter in the soil could be an efficacy strategy to mitigate climate changes and make resilient agroecosystems [22]. As expected, the data showed a positive impact on SOC when O fertilizers are adopted alone or in combination with mineral fertilizer (MO) to replace M fertilizers regardless of all tillage intensities (on average 13.0 and 9.1%, respectively), in agreement with previous studies [23,24]. In a recent piece of research, Rong et al. [25] reported a positive relationship between C input and soil C accumulation, thus meaning that soil carbon accumulation is strongly affected by fertilization programs and the application of organic fertilizers have a positive influence on the increase in SOC. In fact, the application of organic fertilizers supplies the needed substrate for soil microorganisms that are converted into soil organic matter [26].

Similarly, the application of mineral fertilizers may contribute to increasing the SOC in the soil by favoring the accumulation of plant biomass, even if a higher rate of mineral fertilizers rich in nitrogen stimulates the decomposition processes and, thus, determine a progressive depletion of the soil organic matter [27]. This could be the reason why the increase in soil organic matter is higher when organic fertilizers are applied alone (O) compared to their application in combination with mineral fertilizers (MO).

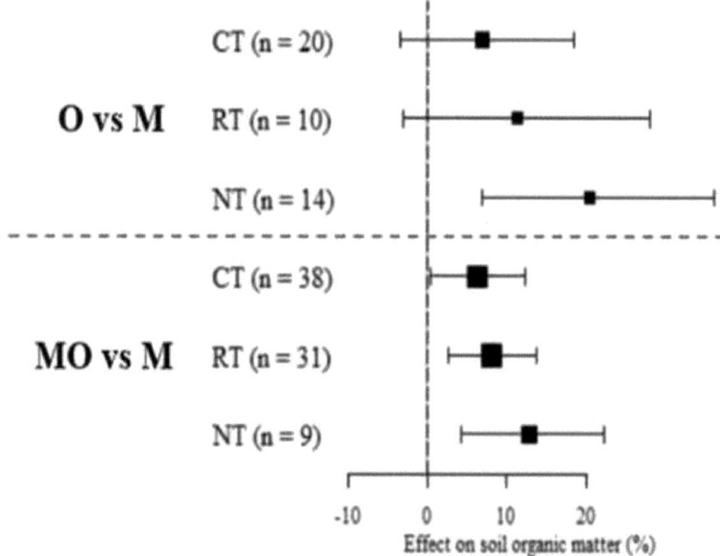

Figure 1. The soil organic matter (SOM) response of organic vs. mineral (O vs. M), and mineral + organic vs. mineral (MO vs. M) under conventional, reduced,

and no-tillage practices (CT, RT, and NT, respectively), expressed as the average effect on soil organic carbon (%). "n" refers to the number of observations for each subgroup. The vertical line represents the null hypothesis [$ln(RR) = 0$]. The squares are the point estimate of effect size. The horizontal lines are the associated 95% confidence interval for the population parameter.

The analysis showed that there is a general trend of increasing SOC concentrations when reducing soil tillage intensity, either using O or MO instead of M fertilizers. Some research reports that conservation tillage practices have been recently supported mainly due to preserving soil health and increasing their fertility [28-29]. Indeed, reduced tillage practices aim to maintain SOC in the topsoil due to reduced contact with soil microorganisms and all related oxidation processes [13,30].

Accordingly, in this study, reduced tillage practices (RT or NT) have increased SOC in comparison with the conventional practice (CT) regardless of the fertilization source adopted (on average 6.6, 9.8, 16.7% in CT, RT, and NT, respectively). Significant effects were detected on SOC when using O fertilization under NT practices (20.6%). Under MO fertilization, the SOC enhancing was about 6.3%, 8.1%, and 12.9% in CT, RT, and NT, respectively. These results provide evidence that conservation agriculture is related to organic nutrient sources combined with conservation tillage practices for improving soil quality through increasing SOC.

The response of fertilizer source comparisons (O vs. M and MO vs. M, respectively) on crop yields subjected to different soil tillage regimes (CT, RT, and NT, respectively) under dry, dry subhumid, and humid climate conditions are presented in **Figure 2**. In agreement with the statement of Hammed et al. [31], climatic conditions could affect the performance of crops subjected to different fertilizer sources and, therefore, improving our understanding of crop responses to fertilizer sources under different climate variations will support practices in climate-smart farming systems by reducing the nutrient loss and improving the nutrient use efficiency.

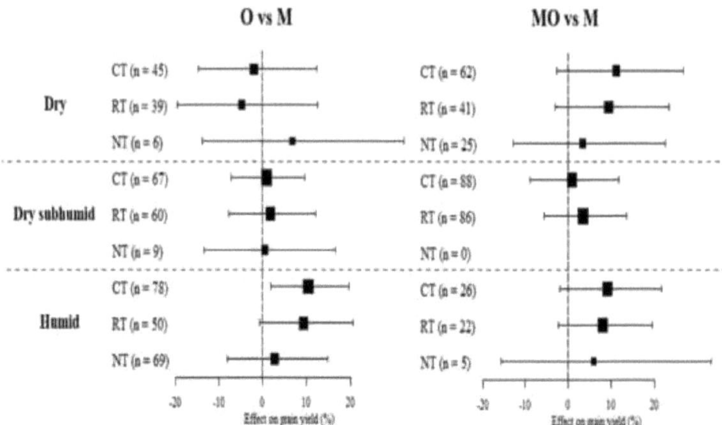

Figure 2. The crop yield response of organic vs. mineral (O vs. M), and mineral + organic vs. mineral (MO vs. M) under conventional, reduced and no-tillage practices (CT, RT, and NT, respectively) in different climatic conditions expressed as the average effect on crop yield (%). "n" refers to the number of observations for each subgroup. The vertical line represents the null hypothesis [$ln(RR) = 0$]. The squares are the point estimate of effect size. The horizontal lines are the associated 95% confidence interval for the population parameter.

In dry climate conditions, O fertilizers applied alone led to negative yields under CT and RT tillage regimes (−2.0 and −4.8%, respectively). Soil tillage may contribute to soil compaction, soil erosion, and excessive organic matter mineralization, that in the long-term period could determine a loss of soil productivity and fertility [32].

Conversely, crop yield under NT resulted higher compared with M fertilizer (6.8%), even if this result should be taken into consideration due to the limited number of observations (**Figure 2**). Several studies reported higher crop yield often reported for NT in dry conditions compared with CT [**9,33,34**].

The higher yield response of crop fertilized with organic fertilizers observed in NT tillage compared to RT and CT soil tillage under dry conditions could be due to a better status of soil microbes. Recently, Schmidt et al. [**35**] observed that the adoption of no-tillage practices shifted the microbial communities toward an increase in stress-tolerant microbes that may support the crop response to adverse agro-environmental conditions. In addition, Wang et al. [**36**] observed that the application of organic fertilizers in arid environmental conditions contributes to improving soil porosity and structure with an improvement on the water use efficiency and soil water infiltration. Conversely, tillage practices may threaten agroecosystems in arid and semi-arid environmental conditions due to negative modifications of soil properties [**37**].

Indeed, intensive tillage frequently repeated over a short period in the same soil can determine a gradual degradation of soil structure and reduce the stability of soil aggregates that lead to reduced soil water availability and soil erosion and compaction [**38**]. In addition, the application of mineral fertilizers and their improper management in dry climates may cause damage to soil fertility with negative consequences for the overall system [**39,40**].

The results showed as the application of organic fertilizers in combination with mineral fertilizers (MO treatment) in dry environments showed higher crop yield when compared with mineral fertilizers applied alone (M treatment), supporting the idea that intensive use of mineral fertilizer may have negative effects on plant growth and yield, but the application of organic fertilizer could alleviate these problems [41].

Similarly, Nouraein et al. [37] observed that the combination of organic fertilizers and balanced mineral fertilizers affects the soil characteristics in terms of enhanced structural stability and increased soil biological activity.

A positive yield trend was observed in dry subhumid and humid environmental conditions, where the adoption of organic fertilizers showed slightly higher crop yield under all soil tillage regimes, even if these effects were mostly not statistically significant.

Significant impact was only detected under humid conditions, where CT was superior compared with other tillage practices (on average 10.4 vs. 6.1%, respectively). Moreover, a positive tendency under all tillage systems in both dry subhumid and humid climate conditions when depending on MO as a nutrient source (on average 16.0 and 7.7%, respectively). However, these trends were also not statistically significant. RT showed higher crop yield than CT in dry subhumid conditions (2.6 vs. 0.9%, respectively), particularly when using MO fertilizers (**Figure 2**), while no observations were available under NT. In agreement with Hijbeek et al. [42], crop yields in humid environmental conditions had more benefits from organic nutrient sources (**Figure 2**).

Under humid environmental conditions, higher crop yield using O could be explained by higher nutrient mineralization and better soil aeration in conventionally tilled soils. According to the findings of Mancinelli et al. [13,43], soil tillage increases the mineralization rate releasing mineral nutrients available for crop nutrition supporting crop yield. However, De Ponti et al. [44] reported that site-characteristics significantly affect the yield gap between organic and mineral fertilizer sources, especially small yield gap was observed in humid environmental conditions. Overall, the results showed that climate conditions have more influence on grain yield when depending on organic sources alone, than combining both organic and inorganic fertilizers, both in comparison with inorganic sources alone. The results highlighted the need to be flexible and environment-specific when considering conservation or reduced tillage and the use of inorganic fertilizers.

Several studies have shown that soil properties and texture are crucial for nutrient availability, especially nitrogen, that is released by means of the mineralization process of organic matter. Overall, soil texture represents a key factor for controlling the soil organic carbon stocks in a specific climatic area [45]. On the other hand, in a previous study on maize crop yield where O fertilizers are compared to MO fertilization management, it has been reported that soil textures had large effects on crop yield [46].

Crop yield responses to organic fertilization and organic plus mineral fertilization against mineral fertilization programs subjected to different tillage practices (CT, RT, and NT) under coarse, medium, and fine soil textures are reported in **Figure 2**.

The analysis showed that in coarse and medium soils, crop yield responses were higher when using O or MO fertilization programs in comparison with M fertilizer applied alone, regardless of the adopted tillage regime (on average 5.8 and 8.4%, respectively). A significant positive impact was detected using O alone in medium soils only under CT and RT (9.5 and 11.2%, respectively), while significant impacts were found in coarse soils using MO sources under both CT and RT (13.4 and 12.7%, respectively).

Similarly, the findings of Lin et al. [47] showed as the adoption of O fertilizer sources under various agricultural practices on crop yield productivity comparable to mineral fertilizer sources, even if under medium-textured soils it was observed a greater crop yield advantage compared with heavy and light-textured soils. In agreement with the results of this study, Allam et al. [48] reported that in comparison with fine soils, O fertilizers alone or combined with M fertilizers under the RT system improved soil structure properties in coarse and medium soils, which leads to considerable yield benefits. The application of organic fertilizers combined with the tillage treatments showed great potential to affect soil microstructure, and, thus, water and nutrient availability for crop growth and yield [49]. Reducing tillage intensity through the application RT or NT tillage regimes in this study led to higher yield in medium soils using MO fertilizers (8.4 and 11.4% in RT and NT, respectively, **Figure 3**).

In fine soils, on the other hand, a negative trend using O fertilizers alone and no trend was detected using MO fertilizers under all tillage practices were detected in this study.

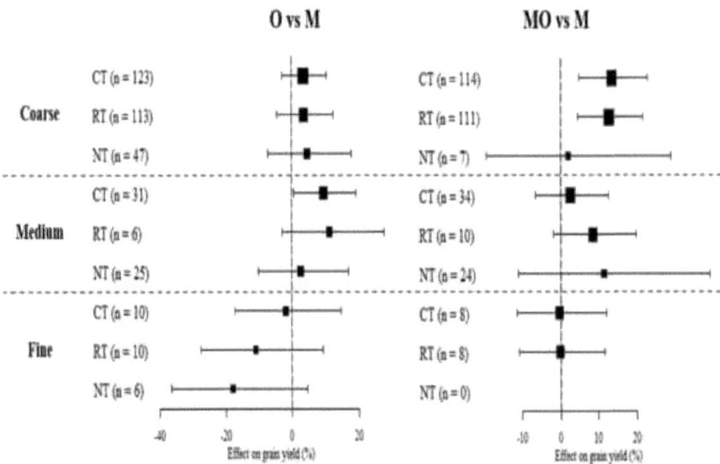

Figure 3. The crop yield response of organic vs. mineral (O vs. M), and mineral + organic vs. mineral (MO vs. M) under conventional, reduced, and no-tillage

practices (CT, RT, and NT, respectively) in different soil types, expressed as the average effect on crop yield (%). "n" refers to the number of observations for each subgroup. The vertical line represents the null hypothesis $[ln(RR) = 0]$. The squares are the point estimate of effect size. The horizontal lines are the associated 95% confidence interval for the population parameter.

In fine soils, a negative impact was detected when using O fertilizer alone under all tillage practices (on average −10.7%), even if the negative impact on crop yield response was not observed when applying O fertilizers combined with M fertilizers (MO treatments, **Figure 3**).

The high clay content in fine-textured soil determines compaction, especially during wet conditions, and thus may initiate serious sealing formation encouraging the adoption of soil tillage to physical improvement to support crop establishment and yield [**17**].

In addition, fine-textured soils have a greater ability to physically protect organic matter in the soil and, therefore, the addition of organic sources by fertilization practices in already organic-rich soils negatively affect crop yield response, as observed by Singh et al. [**50**]. In addition, a positive impact on crop yield response under conservation tillage systems (NT) on coarse and medium-textured soils and negative on fine soils was also reported by Rusinamhodzi et al. [**51**].

The crop yield response due to fertilizer source comparisons (O vs. M and MO vs. M, respectively) subjected to different soil tillage under different crop species are reported in **Figure 3**. Although crop categories could differ in deep-rooted and shallow-rooted crops and thus determining different soil tillage requirements, in this study, all crop species showed a positive trend regarding using O or MO sources under all tillage intensities, except for fiber crops that generally showed a negative impact, even if no significant differences were detected (**Figure 4**).

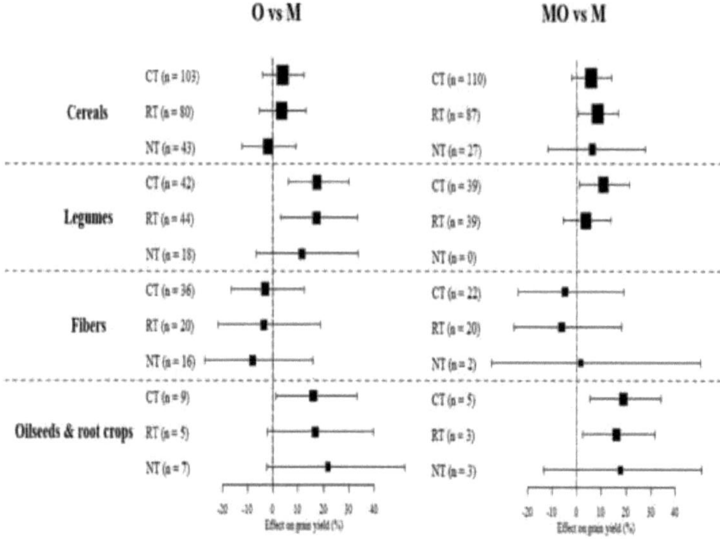

Figure 4. The crop yield response of organic vs. mineral (O vs. M), and mineral + organic vs. mineral (MO vs. M) under conventional, reduced and no-tillage practices (CT, RT, and NT, respectively) in different crop categories, expressed as the average effect on crop yield (%). "n" refers to the number of observations for each subgroup. The vertical line represents the null hypothesis $[ln(RR) = 0]$. The squares are the point estimate of effect size. The horizontal lines are the associated 95% confidence interval for the population parameter.

A significant positive impact was reported for cereals using MO under the RT system (8.4%), while a negative trend was observed for cereals only using O fertilizer sources under NT (−2.0%). The application of MO seems to be the best option for higher grain yield for cereals under RT. The combined application of inorganic (M) and organic (O) fertilizers has led to increased cereals yield compared to M or O fertilizers alone [46,52].

Similarly, Campiglia et al. [53] observed that durum wheat yield was similar among conventional and reduced tillage, especially in organic cultivation, and therefore concluded that the adoption of reduced tillage practices for durum wheat cultivation was preferable because it supported the reduction of intensive tillage.

Under the O fertilizers, the legumes yield response was significantly greater under both CT and RT regimes (on average 17.5 and 17.4%, respectively), while no significant response was observed under NT (**Figure 4**).

Similarly, Zingore et al. [54] showed a higher performance of soybean crops when organically fertilized compared with mineral fertilization. The great yield response to organic fertilization was mainly attributed to the enhanced soil conditions needed for legume performance [55].

In addition, legumes can fix atmospheric N; thus, it is not a limiting nutrient for them, and consequently, they are more tolerant to the mineralization process in comparison with cereals crops. The application of MO fertilizers could give significant benefits for legumes only under CT techniques, but no significant response was found under RT; no observations were collected from experiments that evaluated legumes using MO and NT in our study. In pea crops, Faligowska et al. [56] observed higher grain yield in conventionally tilled soil compared to no-tillage, suggesting that these differences could be related to different biological and physical properties between the different tillage regimes.

The previously mentioned study evaluated the impact of using O versus M on yields under various agricultural practices [47]; it has also been reported that legume crops performed significantly better when using an O source, with a positive pattern similar to this study was noticed over all crop species. Similarly, under organic farming managements, Cooper et al. [57] have reported that legumes were the best performed when reducing tillage intensities.

The yield response of fiber crops was positive only with MO under NT (1.8%), even if this result is limited due to a very small sample size (n = 2), in accordance with the findings of Idowu et al. [58] that reported several benefits of reduced tillage in cotton crop. Moreover, oilseeds and root crops showed a significant increase in grain yield using O with CT (21.8%) and MO with CT and RT (18.8 and 16.2%, respectively). A recent study showed as the seed yield of oilseed rape slightly varied according to the soil tillage regime suggesting the adoption of reduced tillage practices for this crop in order to improve economic benefits for the farmers [59].

The impact of using O or MO instead of M fertilizers under different tillage systems on crop yield response under two rainfed and irrigated conditions is reported in **Figure 5**. Water represents one of the main limiting factors for crop production, and a rational fertilization strategy should consider soil water availability to improve crop yield in a sustainable way, according to the findings of Liu et al. [60].

The analysis showed a positive trend under all tillage practices under both irrigated and rainfed systems, even if the only significant impact was detected under rainfed conditions using MO and CT (10.0%). Celik et al. [61] reported improved soil physical properties when organic materials were added to the soil, while Nyamangara et al. [62] observed a better soil water retention capacity when cattle manure was applied to agricultural soils. It is conceivable that the application of organic fertilizer applied alone or in combination with mineral fertilizer benefits rainfed agroecosystems because it maintains the soil water storage balance, supporting an increased availability of soil water for crop growth and yield compared to mineral fertilization programs [**63, 64**].

Although no differences were detected among the tillage regimes, the lower crop yield response observed in NT than CT and RT could be a result of high variability and unequal distribution of rainfalls that cause high loss of nutrients, especially nitrogen, mineralized from O fertilizer sources [**53**]. The study suggests that under rainfed conditions, when depending on O or MO nutrient sources, RT is more suitable than NT for ensuring higher crop yield.

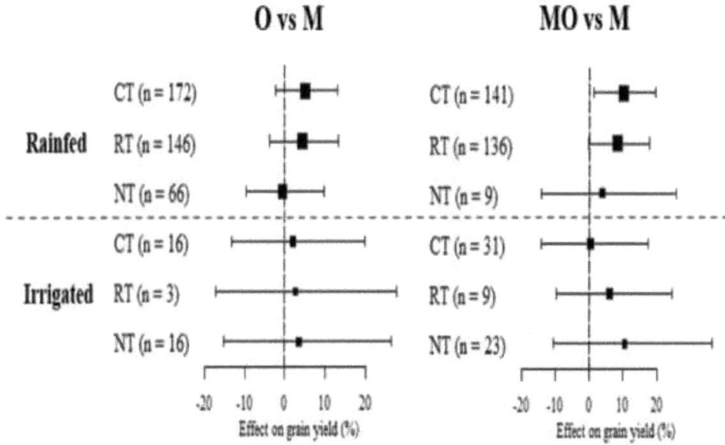

Figure 5. The crop yield response of organic vs. mineral (O vs. M), and mineral + organic vs. mineral (MO vs. M) under conventional, reduced and no-tillage practices (CT, RT, and NT, respectively) in different irrigation regimes, expressed as the average effect on crop yield (%). "n" refers to the number of observations for each subgroup. The vertical line represents the null hypothesis [$ln(RR) = 0$]. The squares are the point estimate of effect size. The horizontal lines are the associated 95% confidence interval for the population parameter.

Under different irrigation management, it seems that RT and NT tillage regimes led to higher productivity than CT (on average 2.4, 4.3, and 6.9 in NT, RT, and CT, respectively), especially when organic fertilizers were applied in combination with mineral fertilizer (**Figure 5**). Additional water supplied through irrigation enhanced plant N uptake, especially under conservation than conventional tillage systems [**28**]. It also highlights that additional water supply could be a good strategy when depending on MO under conservation tillage practices.

The difference between mineral and organic fertilizers is their composition and functions. Mineral fertilizers provide large amounts of nutrients that plants need to grow strong. Organic resources contain organic carbon which is an essential ingredient for healthy soil.

To be able to create the optimal NPK proportion, the right amount of these nutritional elements is of high importance. NPK stands for the elements nitrogen (N), phosphorus (P) and potassium (K), which are needed for a solid base. In this article, the difference between organic and mineral fertilizers, also called chemical fertilizers, is explained by our fertilizer specialist.

Organic fertilizers

Organic fertilizers consist of natural materials, such as bacteria, molds, insects, worms and other organisms. The soil live is stimulated by these natural materials.

Organic fertilizers often need to be converted by certain organisms in the soil, in order to become available to the plant as nutrition. The moment, amount and order in which the elements are taken in, are decided by the plant itself. Organic fertilizers indirectly ensure the plant's resistance for plagues, viruses and diseases.

For some cultivators, the use of organic fertilizers brings along a disadvantage; the product can smell. This because they're made of natural materials of plant or animal origin. Apart from the disadvantages, organic fertilizers also have many advantages. Our fertilizer specialist highlights some of the advantages.

Mineral fertilizers

Mineral fertilizers, also known as chemical fertilizers, are not fully built on natural materials. This, because mineral fertilizers emerge after a chemical process. Nonetheless, the materials that can be found in these fertilizers, can also be found in the natural environment.

All elements that are present in a mineral fertilizer can be taken in by the plant instantly, since no conversion is needed. This is seen as an advantage, because the fertilizer is not dependent on the soil live. Besides, these fertilizers are easily soluble and therefore easy to add to the plant. In addition, the nutritional scheme can be composed in line with the plant's needs and is therefore always in line with the cultivator's wishes.

As opposed to making use of organic fertilizers, the use of mineral fertilizers gives the plant the opportunity to 'choose' one or more element which is needed at that specific moment. With mineral fertilizers the possibility for a more steered/aimed fertilization is given. Therefore, it is possible to meet all of the plant's needs, which are calculated in the nutritional scheme. The crop's 'choice' is no longer necessary.

The soil live is not contributed to by mineral fertilizers, which can be seen as a disadvantage. Also, the amount of sodium in mineral fertilizers is higher than in organic fertilizers. Sodium could pile up and cause damage in the roots, if mineral fertilizers aren't added to the plant correctly. This is definitely something that should be paid attention to when making use of mineral fertilizers.

The right choice of fertilizer for your crop, is dependent on several factors. The type of substrate, nutritional need and sensitivity of the crop, and the way in which you like to add fertilizers to the crop, are factors that should be taken into account when making a decision.

In the process of making a choice between the two types of fertilizers, it is recommended to be advised by one of our fertilizer specialists.

Five important facts about fertilizer

Fact 1: Fertilizers do not deplete the soil

Fertilizers are key to rejuvenating the soil by providing nutrients the plants need to grow healthily.

In nature there are 17 critical plant nutrients: The macronutrients nitrogen, phosphorus, potassium, calcium, sulphur, magnesium, oxygen, hydrogen, carbon, and the micronutrients iron, boron, chlorine, manganese, zinc, copper, molybdenum and nickel.

When crops are harvested, the nutrients follow the crop. Important nutrients are therefore removed from the soil. Often the soil is not able to replenish all the nutrients by itself, that is where fertilizers supply the nutrients that are lacking.

To keep up with the world's rising population, higher crop yields are essential. In the US alone, the average corn yields have more than doubled since 1968, through more effective farming.

Both organic and mineral fertilizers can be used to replenish the soil. The nutritional content of organic fertilizers is low compared to mineral fertilizers, which are concentrated and have a strictly controlled nutrient content.

Fact 2: Fertilizers are made of natural elements

All the nutrients contained in different fertilizers are found in nature. The most common sources of nutrients in mineral fertilizers are nitrogen, potassium and phosphate.

Nitrogen originates from the air. The most common process in nitrogen fertilizer manufacturing is to create ammonia from a mixture of nitrogen from the air and hydrogen from natural gas.

Air consists of 78 percent nitrogen, but plants cannot get the nitrogen needed directly from the air – they need to take it up through their roots from the soil. Potassium is sourced from old sea and lake beds formed millions of years ago.

Potassium fertilizers are based on naturally occurring potassium chloride. This is somewhat similar to table salt – sodium chloride.

The ash from burning wood or straw is high in potassium, this is where the name 'potash' originates.

Since potassium sources are often located far below the soil surface (1-2km depth), plant roots are unable to reach them naturally.

The world's biggest potassium producers are Canada, Russia, Belarus and China.

Phosphate is sourced from insoluble calcium phosphate rocks – often referred to as "rock phosphate". In this form it is not available to plants. Rock phosphate is made available for the plant usually through a chemical process to create plant friendly fertilizers.

China, Russia and Morocco have some of the world's largest deposits of phosphate rock.

Nitrogen (N), phosphate (P), and potassium (K) can also be combined to form NPK compound fertilizers, that provides the crop with the 3 major nutrients at the same time.

The alternative to mineral fertilizers is organic fertilizers which are based on materials with a biological origin. These include animal wastes, crop residues, compost, biosolids and more.

Without fertilizers, the soil would be depleted and therefore plants would be particularly difficult to grow. They cannot survive on water alone, and nor can we.

Fact 3: Fertilizers are not the same as pesticides

Pesticides are synthetic or natural chemicals used to control pests. Pesticide is a commonly used term for all crop protection chemicals, which also include fungicides that control fungal diseases, herbicides that control weeds.

Fertilizers, on the other hand, supply natural nutrients to make crops grow.

The role of fertilizers is to increase yield and ensure healthy produce by supplying the right balance of nutrients to the soil.

"Without fertilizers, the soil would be depleted and therefore plants would be particularly difficult to grow. They cannot survive on water alone, and nor can we. If we want good nutritious food, plants need nutrition and that makes our food much more enjoyable," says Barry Bull, plant nutrition consultant.

Fact 4: Fertilizers do not alter the plants we eat

Fertilizers do not alter the DNA of crops. Instead they improve the growth and quality of the crop by adding important nutrients.

The amount of nutrients added is chosen by the farmer after analysing the soil and determining the requirements of individual crops.

Fertilizing in the correct way can have a great impact on crops' yield, appearance and nutritional value.

Fertilizer adds nutritional value: To the right, tomato plant with nitrogen deficiency, to the left, tomato plant with optimum nutritional balance.

Fact 5: Fertilizers do not make you ill

Eating crops from a fertilized field, or meat from animals that have grazed on a fertilized pasture, does not pose any health risks for animals or humans.

On the contrary, the nutrients in the fertilizer required for crop growth, are the same nutrients required for human growth and development. It is a fact that approximately half of the world's population today has food on the table due to fertilizers.

Careful fertilizing is key to increasing crop yields on existing farmland, which in turn helps combat disorders caused by malnutrition.

In countries where specific nutrient deficiencies are a problem, fortifying fertilizers with the relevant micronutrients have also helped better the health of large populations.

Zinc and selenium are two examples of minerals that have been successfully applied to fertilizers to combat deficiencies in large populations.

Mineral fertilizer by BAC

There are two main types of fertilizer: organic fertilizer and ***mineral fertilizer***. Each type of fertilizer is created in its own way, and each offers its own unique advantages.

Organic fertilizer is made from plants or animals, while mineral fertilizer is created through chemical processes; however, the components present in mineral fertilizer can also be found naturally in the environment. Mineral fertilizers are composed of three main nutrients: nitrogen, phosphate, and potassium. The amount of each present in any given mineral fertilizer will be indicated by the NPK ratio on the product packaging.

Other nutrients may also be included in small amounts, such as calcium, magnesium, or iron. Mineral fertilizer provides these nutrients to plants in a pure form. This means the nutrients from mineral fertilizer are instantly available to the plant, and therefore can be absorbed by the plant straightaway. This differs from organic fertilizer which requires micro-organisms in the soil to first breakdown the organic matter in order to release the fertiliser nutrients. Therefore, mineral fertilizer is fast-acting, while organic fertilizer is long-lasting. Find out more about mineral fertilizer

Mineral Fertilizer, Application

Mineral fertilizer is typically soluble and can be added to plants while watering. Since the nutrients from mineral fertilizers are instantly available to plants, **mineral fertilizer** can be used to aid a plant's development at specific moments.

Alongside being generally fast-acting, mineral fertilizers offer the added benefit of being able to supply the exact nutrients a plant needs, in the moment it needs them. This means mineral fertilizer can also be used to address nutrient deficiencies in plants as and when they occur.

Mineral Fertilizer, Risks

While mineral fertilizer offers many benefits, it must be applied with a certain amount of caution. If mineral fertilizer is applied incorrectly salts can accumulate in the soil and damage a plant's root system leading to fertilizer burn.

What is more, because mineral fertilizers are soluble they have a tendency to leach into the groundwater and away from the plant. It is therefore important to add mineral fertilizer at the right time. For best results, follow BAC's plant feeding .

Mineral Fertilizer from BAC

Mineral fertilizer from BAC has been carefully researched to ensure the exact balance of nutrients a plant needs to flourish. BAC's mineral fertilizers offer precise fertilisation of plants with fertilizer that contains all the necessary nutrient elements. BAC has developed several mineral fertilizers, with products suited to a variety of growing mediums.

In the long-term biological–geological co-evolution, the life activities of surface microorganisms play an important role in a series of surface geochemical processes, such as rock weathering, soil formation and evolution, which directly or indirectly affect the growth of various soil organisms, including the massive formation of terrestrial vegetation.

It can be said that the growth of surface plants depends to a large extent on the mineral nutrients released by soil microbial–mineral interactions. With population growth and human science and technology progress, the yield of crops increased significantly. The nutrients provided by the weathering of rocks under natural conditions was no longer able to meet the needs of the normal growth of plants, prompting the discovery and utilization of chemical fertilizers in the 18th century, and their rapid development as an important guarantee for improving crop yield and quality.

Intense agricultural practices, especially the indiscriminate utilization of synthetic compounds (N, P, and K) has been increasing steadily in most parts of the world. Intense agricultural practices combined with approaches such as the reclamation of degraded soils to enhance food productivity, have created repercussions beyond human control/comprehension. The degradation of fertile soils, attributable to intense agricultural practices, is a major hurdle for food security and sustainable agricultural development.

As degraded soils feature relatively lower levels of essential soil nutrients, active organic carbon, low or high pH coupled with limited soil enzyme activity usually generate low yields and nutritionally subpar produce. Furthermore, excess chemical fertilizers impact the rhizosphere's microbial communities adversely, disrupting their normal soil functions (nutrient cycling, organic matter creation, soil nutrient improvement).

However, due to rapid developments in industry, agriculture, and population growth, large areas of fertile soils were either degraded or converted into non-agricultural activities [65], and many areas of newly reclaimed soil (most of which are poor soils), have been adapted for plant cultivation. Many reclaimed soils, along with degraded soils, are characterized by sufficient contents of nutrients (K, Ca, and Mg) but remain in indiscerptible chemical states, are not absorbed by the plants, and often present as soil minerals. In addition, the long-term use of chemical fertilizer leads to the increasingly evident problem of heavy metal pollution in cultivated soil.

Despite the innovations and modern approaches in recent agricultural practices, the majority of the global agriculture sector still depends on conventional practices; thus, it suffers from sustainability and fertility issues. Moreover, soil fertility continues to decline due to the increase of multiple cropping practices and increasing agricultural yields through the misuse of chemical fertilizers [66].

Hence, measures to improve these soils are urgently required, as global agriculture is being significantly hampered by various yield-limiting factors (increased disease occurrences, pest and insect invasions, erratic weather). Furthermore, agricultural activities are controversially deemed a major CO_2 contributor to damage to the atmosphere, driving global warming.

Excessive use of chemical fertilizers not only increases production costs, but also causes problems such as energy depletion, resource scarcity, and environmental pollution as well as compromising food safety, which have become increasingly prominent challenges in recent years. Against the current background of the vigorous development of ecological agriculture and environmental protection, the role of organic matter returning to the field and the development of organic fertilizers on this basis has received renewed attention in agricultural production. Adding organic matter to cultivated soil increases the soil's organic content, regulates the soil's physical and chemical properties, and improves the quality of agricultural products, but a large amount of organic matter in cultivated soil inevitably increases cultivated soil respiration. Facing the global greenhouse effect, the negative impact of organic matter returning to farmland should not be ignored.

Minerals are also an important component of the soil; they are the skeleton of the soil and the source of mineral elements. Minerals play an important role in the improvement of soil's physical and chemical properties and the growth and metabolism of microorganisms. However, the beneficial effects of the use of appropriate minerals in the soil have long been neglected. In addition, our research has found that forms of mineral weathering, such as silicate weathering, are often accompanied by the formation of secondary carbonate minerals in the process of biological weathering, which undoubtedly increases the potential of cultivated soil carbon sinks [67- 70].

Further studies have shown that secondary minerals formed with mineral weathering have a good remediation effect on heavy metal pollution [71, 72]. In addition to the formation of secondary minerals, the cations released by mineral weathering can also combine with the soil's organic complexes through co-precipitation, which in turn mediates the formation of soil aggregates, preserving soil organic carbon, thereby reducing the potential of soil carbon depletion. Therefore, bio-organic mineral fertilizers (BOMFs) have a positive impact on agriculture, soil health and the ecological environment [73- 78].

In this review, we emphasize the application urgency, production methods and ecological effects of BOMFs, and point out the future research direction of BOMFs, hoping to provide a new vision for the development of sustainable agriculture.

By the end of 2020, the total global amount of fertilizer (N, P and K) reached more than 200 million tons per year [**79**]. This is alarming as fertilizers have severe adverse effects on soils as well as on surrounding ecosystems. In addition to this, the use of pesticides to prevent and control plant diseases is accelerating at a much more perilous rate. China (1,773,676 t), the United states of America (407,779 t), and Brazil (377,176 t) occupy the top three places in the world for pesticide utilization as of 2019, according to FAO.

However, concerning the consumption rate (kg/hectare), Maldives (52.6), Trinidad and Tobago (24.9) and Costa Rica (22.5) are the top three consumers of pesticides despite their smaller areas of agriculture [**80**]. Long-term conventional farming has far reaching consequences for human health and the environment [**66,81-** 83].

As a response to the many hurdles, contemporary and very recent research has demonstrated that soil quality can be improved by applying organic matter, organic fertilizer (OF), dust, and mineral powder (MP), and/or by growing pasture legumes as green manure [**84-** 86]. The natural breakdown of many minerals due to microbial metabolic activities, combined with abiotic factors innate to a given ecosystem, provide sustenance to plant growth [**71,87**].

It is a widely acknowledged fact that organic matter is a crucial component for improving the physical properties of soil, and that returning organic matter to soil can increase the content of soils' active organic carbon (AOC) and improve soil vitality. However, organic matter alone, which lacks adequate accessible nutrients (for plants), exerts slow and variable effects on crop growth [88].

Although applying chemical fertilizers (CFs) can promote rapid crop growth, extensive CF use has reduced soil quality worldwide [89]. Utilizing and/or enhancing the mineral weathering potential of many microbial groups (bacteria and fungi) coupled with mineral ore (rock phosphate, feldspar, etc.) application has been reported to exert extremely positive effects on soil health and fertility, an approach devoid of environmental concerns [77,87,90,91].

Hence, compound fertilizers composed of suitable microbes, organic matter, and raw mineral powders, along with low quantities of CF, offer the potential to achieve sustainability goals under the prevailing environmental scenario [74,92,93].

Contemporary research shows that potassium-containing minerals transformed by microorganisms in soil can promote crop growth [94- 97]. For example, *Paenibacillus mucilaginosus* (also known as silicate bacterial in China) is an important species, widely used in microbial fertilizers in China [97- 100].

Lian et al. (2002 and 2020) explored the potassium-releasing effect and mechanism of bacteria, and proposed the comprehensive effect of the potassium-releasing mechanism by bacteria. The potassium release effect of silicate bacteria on aluminosilicate-containing potassium minerals is based on the formation of a bacteria–mineral complex through the combined effects of acidolysis, chelation, dissolution, and active absorption, which promote each other and together lead to the gradual release of potassium ions in minerals.

Further, the micro-environment formed by microbial extracellular secretions and microbial–mineral interactions (bacteria–mineral complex) is closely related to the release of potassium ions from mineral weathering [**73**- 75]. These findings helped to further the understanding of the mechanism of weathering potassium-bearing minerals to improve soil potassium utilization rates through silicate bacteria.

In some areas, Chinese farmers have used potassium-containing mineral powder, phosphate rock powder, agricultural by-products, and organic wastes to make compost [**69,101,102**]. This method of fertilizing fields by supplementing potassium and phosphorus is simple but varies greatly in its effect on crops. In recent years, some scholars have also begun to try similar methods to obtain organic potassium phosphate fertilizers using different organic materials and minerals, supplemented with microbial agents [**72,78,103**- 105].

The results indicate that fermentation with a microbial agent greatly increase the soluble content of nutrients (K and P), suggesting that microbial fermentation can efficiently transform low-grade mineral rocks. Furthermore, potassium-containing rocks and auxiliary materials can be converted into multifunctional compound organic mineral fertilizers, which can not only supplement the nutrients needed for crop growth and improve its quality, but also repair inferior soil that has been severely degraded due to the excessive application of chemical fertilizers [**69,77,92**].

In summary, BOMFs are fertilizers prepared with minimally processed components (chemical agents such as urea, phosphates, etc.). BOMFs mostly comprise mineral sources and animal excreta (cow dung, chicken feces, etc.) as bio-organic components and differ significantly from conventional organic fertilizers or bio-fertilizers in their composition and resultant effects. The components and basic differences with conventional chemical fertilizers are presented in **Table 1**.

Table 1. Differences in compositional and functional attributes between BOMFs and conventional synthetic fertilizers.

Different Distinguishing Components	BOMFs	Conventional Fertilizer
Mineral source (P, K, Ca, B, Mg, etc.)	Natural	Chemical
Mineral (dust or powder)	✓	x
Microbial agents	✓	x
Plant growth-promoting microbes	✓	x
Plant protecting microbes	✓	x
Organic carbon	Natural (agri. waste)	x
Nitrogen source	Natural (animal waste)	Chemical (urea)
Inorganic additives	xx	Chemical additives (P, K, etc.)
Negative impact environment	x	✓

Different Distinguishing Components	BOMFs	Conventional Fertilizer
Known CO_2 sequestration potential	✓	x
Positively influences soil parameters	✓	x

Note: ✓ means contained, x means excluded, xx means excluded unless otherwise noted (adding a small amount of chemical fertilizer).

Additionally, the organic component serves two functions: as a carrier for the microflora and as an active soil carbon source, further improving soil fertility. The most noticeable feature of BOMFs is the presence of a natural, unprocessed mineral source instead of chemical agents for supplementing macro-elements (phosphate, potassium, calcium, magnesium, sulfur, boron, etc.).

Phosphate (P), the most widely applied non-nitrogen fertilizer, features limited bioavailability in its raw form (rock phosphate or phosphorite), needing further chemical processing before its application to soil. This can be managed with BOMFs, where rock phosphate itself is used as P source that is gradually weathered by the action of innate/or added microflora in BOMFs. Besides P, many other macro-elements can be supplemented as their raw sources (feldspar, muscovite-potassium; dolomite-magnesium; calcite-calcium; gypsum-sulfur) before high-temperature fermentation. Furthermore, BOMFs can be prepared with widely available materials, reducing the manufacturing process as well as the financial pressure on the farmer to obtain healthy produce from the fields. BOMFs could also be much more effective in soils lacking pedogenic parent minerals (such as karst soils in Southwest China, sodic soils in south Asian countries, and sandy soils in Africa) as the mineral component can be gradually weathered by microbial community-enhancing soil agricultural attributes.

Thus, the organic mineral fertilizer obtained by comprehensively using organic waste and cheap low-grade mineral powders (such as potash mineral powder and phosphate rock powder, etc.) and adopting microbial engineering fermentation technology can partially replace chemical fertilizers to promote crop growth, improve low-quality soil, and exert a positive ecological effect (**Figure 6**).

Figure 6. Simplified production process of BOMFs.

The preparation of BOMFs is less complex and chiefly involves simple biological and physical methods (**Figure 6**). The components or the general make-up can vary greatly depending upon the soil type, location (e.g., tropical, sub-tropical, or temperate), physico-chemical traits and deficiencies, etc. For example, the organic component can be dependent on the most abundantly or conveniently available source, such as agricultural waste, animal manure and/or compost [**67,77,92**].

Animal manure (chicken excreta or cattle dung) or vermicompost can be used with little processing, and without the need for high-end processes. Furthermore, the mineral component is generally a mineral ore (commonly low-grade) composed of different macronutrient elements.

Minerals such as phosphorite, feldspar, gypsum, basalt, etc., containing macronutrients (such as P, K, Ca, Mg, S, B), are common mineral additives. Thus, the compositional make-up of the components and the produced BOMF can greatly vary. The bio-organics' and minerals' basic composition is provided in **Table 2**.

Table 2. Basic compositional make-up of different components employed in the production of BOMFs.

Ingredient/Component	Major Nutrients	Nutrient Availability
Mineral		
i. Phosphorite or rock phosphate	Phosphorous (P_2O_5) 16–17%	Slow
ii. Feldspar	Potassium (K_2O) 2–12%	Very slow
iii. Gypsum [a]	Sulphur ($CaSO_4$ 12–16%)	Slow
iv. Basalt [b]	Iron (FeO 5–14%) and magnesium (MgO 5–12%)	Slow
v. Calcite or dolomite	Calcium (CaO ~12%) Magnesium (MgO ~20%)	Very slow
Bio-organics		
i. Animal manure [c] (cattle dung and poultry excreta)	Nitrogen (~14%), potassium (~20%), and Phosphorous (~25%)	Quick
ii.	Potassium, nitrogen, and	Quick

Ingredient/Component	Major Nutrients	Nutrient Availability
Compost [d]	phosphorous	

(Source: [102], [a] [106]; [b] [107]; [c] [108]; [d] [109]).

BOMFs significantly increase the organic content in soils as they provide a stabilizing balance between mineral, biological, and organic components. The microbial community innate to the organic material weathers the powdered ore slowly but steadily, creating a stable source of nutrients and improving the soil parameters, especially if active bacterial agents are applied as additives.

According to the typical fertilization situation, the amount of chemical fertilizer used is about 4.5 t/hm^2, and the input cost is 4800–6000 RMB. Although the organic content of organic fertilizer is higher, the pure nutrient content is much lower than that of chemical fertilizer. The use amount of organic fertilizer converted according to the nutrient content of chemical fertilizer is about 22.5 t/hm^2. Taking into account the transportation cost and labor cost, the cost of using organic fertilizer needs to be increased by about 2500 RMB [110], Therefore, the use of commercial organic fertilizers needs to be subsidized by the government, otherwise it will be difficult to use on a large scale.

Compared with the former two, the usage cost of BOMFs is close to that of high-grade organic fertilizer. Since the use of mineral powder and organic matter is mainly based on the principle of using local resources (mineral and organic resources), the cost of using BOMFs will include a significant amount of room for reduction. The following mainly explains the advantages of using BOMFs from the perspective of improving soil's physical and chemical properties and increasing soil carbon sinks.

Ecological Effects of BOMFs

The application of BOMFs in agriculture can improve soil's physical and chemical properties, regulate soil ecological communities, and reduce environmental pollution. Beneficial rhizosphere bacteria play a critical role in the agricultural ecosystem i.e., by preserving soil quality and maintaining soil fertility and crop productivity [111]. An appropriate community structure, rich diversity, and high microbial activity are significant factors in maintaining soil ecosystems and productivity [112].

Impacted Microbial Functions and Plant Growth Promotion

Years of different types of fertilizer application (including NPK, manure, and cattle slurry) demonstrated statistically significant differences in the content and quality of humic substances, soil reaction and available nutrient content.

Based on principal component analysis, fertilizers are divided into two categories: (1) NPK treatment—higher acidity, lower soil organic carbon and humic substances content, higher content of available nutrients; and (2) organic manure treatments—lower acidity, higher soil organic carbon and humic substances content, lower content of available nutrients. Organic manures are known to improve microbial community stability and enhance associated functions [113].

Furthermore, long-term NPK application without organic input can accelerate humus mineralization and soil degradation due to soil acidification, with several negative consequences, such as low nutrient availability, nitrogen leaching, high toxic element availability, fewer sources and energy for microorganisms' activity etc. [**114-** 116].

Amendment through the use of organic fertilizers (manure and cattle slurry) and BOMFs can help to achieve stable yields in the long term while maintaining optimal soil properties. The total microbial diversity and density significantly increase along with the microbial community-associated functions (soil enzyme activity, organic carbon and total available nutrients such as N, P, and K) [**117-120**]. BOMFs also enhance the disease suppression potential of soils by maintaining stable rhizosphere microbial communities [**121**].

Restoration of Soil Fertility and Organic Carbon

The loss of fertility due to intense agricultural practices in the current scenario is of immense significance. Enhancing fertility in intense farms or in non-arable lands is a possible approach to increasing productivity, thus increasing food production without further burdening natural resources.

The unique combination of BOMFs with the supplementation of natural organic matter, raw natural ores instead of synthetic fertilizers, and microbial assemblages capable of weathering, metabolizing and nourishing, could create a stable agricultural microenvironment.

Sun et al. (2019) reported an increase in the soil water holding capacity and agronomic characteristics along with an increase in the biomass of Chinese *cabbage* when fertilized with BOMFs (in this case, potassic rock and organic waste) [**69**].

Additionally, the concentration of inorganic carbon (largely in the form of HCO_3^-) in surface run-off water treated by BOMFs was higher than in the other treatments, establishing carbon sequestration's potential. The combined addition of dolomite and K-feldspar mineral powder to soil has been shown to be effective at improving the plant growth parameters as well as carbon capturing. Plants and microorganisms weather minerals by secreting organic acids, which release cat ions, such as Ca^{2+} and Mg^{2+}. The released cat ions manage atmospheric CO_2 by forming carbonates under a series of biological actions [67- 69,87].

In addition, the cat ions released by mineral weathering are often combined with organic matter, which is sequestered in secondary carbonate minerals when carbonate minerals are formed. These sequestered organic carbons are difficult to decompose and utilize for organisms, thereby increasing soil organic carbon. In one study, the carbon content of the soil was enhanced as mineral weathering accelerated fixation of organic and inorganic carbon. Moreover, the available potassium content was also increased when K-feldspar was added [68].

Facing the global greenhouse effect and cultivated soil degradation, adding moderate amounts of carbonate, phosphorus, and silicate minerals into soil is a sustainable approach through which to accelerate atmospheric CO_2 fixation and improve the nutrient content of soil.

Remediation of Degraded Soils: Cultivation of Specific (Resilient) Crops

Soil fertility degradation is a serious and continuously increasing threat to agricultural soil productivity. Thanks to the use of BOMFs, resilient crops such as *purslane* have shown promising potential for reducing the negative effects of intense cultivation under the influence of chemical fertilizers, as observed in our research [77].

Yu et al. [103] and Xiao et al. [67] reported that the heavy use of chemical fertilizer causes increasing soil and environmental crises, and amendment through the use of organic components and biological agent increases soil health. The major functions, such as pH, organic carbon content, microbial biomass, urease activity, and available potassium content were influenced when BOMFs were used. The cultivation of crops such as *amaranth* and *purslane* [77] exhibited the greater positive impact of BOMFs on soils and environments.

In the case of both *amaranth* and *purslane*, the above-ground biomass and nutritional parameters of the plant were improved, suggesting the positive impact of BOMFs on agricultural soil as well as crops. It should be further noted that BOMFs are also affective in remediation of soil pollution, as reported by Chen et al. [104]. In pot experiments simulating karst mountainous area and heavy metal contamination (cadmium) using *pakchoi cabbage* (*Brassica rapa chinensis*), the results demonstrated that the heavy metal accumulation in BOMF-treated plants was the lowest, suggesting the effectiveness of mineral–microbial combinations.

In summary, the application of BOMFs in farmland can improve the physical and chemical properties and structure of the soil [122,123], regulate soil microbial community composition [82,123], and promote the healthy and robust growth of crops by increasing the richness and uniformity of beneficial microbial communities [124] (**Figure 7**).

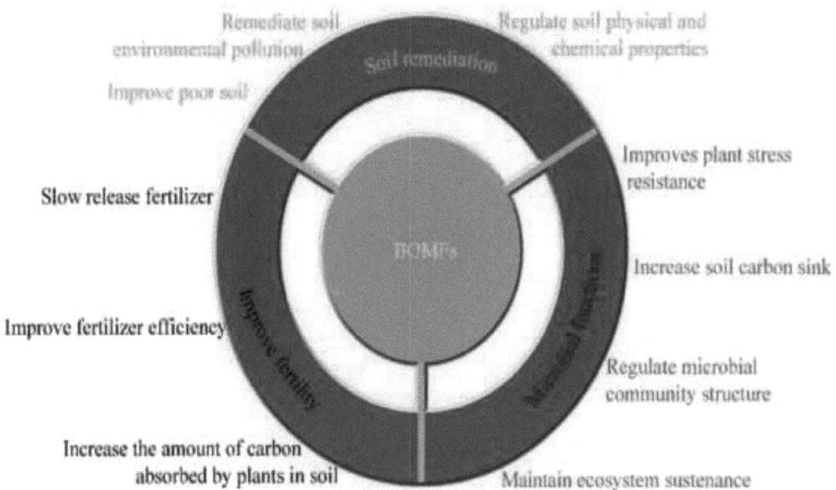

Figure 7. Ecological effect of BOMFs application.

In addition, the improvement of agricultural soil can improve the agricultural ecological environment, improve the quality of agricultural products, increase economic income, and promote the prosperity of people in rural areas so that they can live and work in peace and contentment.

There is no doubt that organic fertilizers, such as manures, compost, and bone meal, are generated directly from plant or animal sources.

Whereas commercial or synthetic fertilizers, such as ammonium sulphate or ammonium phosphate, are derived from naturally existing mineral deposits.

Only a few nutrients, including nitrogen, phosphorus, potassium, sulphur, and, on rare occasions, micronutrients, are usually contained in inorganic fertilizers, either alone or in combination.

These nutrients have been converted into a form that plants can utilize.

This blog will mainly discuss the importance of organic fertilizers to plants.

But before proceeding ahead, let us know what basically is the definition of organic fertilizers.

What is Organic Fertilizer?

"Organic fertilizers are mineral sources that are found in nature and contain a sufficient amount of plant nutrients."

Organic fertilizers are made from plant-derived resources, which might be fresh or dried plant material.

The nutritional content of organic fertilizers varies widely depending on the source material, with rapidly biodegradable compounds serving as suitable nutrient sources.

The organic carbon content of an organic fertilizer may be as critical, if not more so, than the nitrogen and phosphorus concentrations.

Organic fertilizers, which promote secondary productivity and mineralize nutrients to raise primary productivity, produce increases in heterotrophic bacterial biomass.

Farmers, landscapers, and gardeners will find these fertilizers to be incredibly cost-effective.

Homeowners can utilize natural fertilizers to improve the performance of their plants. Organic fertilizers are simple to use and fully safe to utilize.

These natural products are easy to use and do not require the use of costly protective clothing or equipment.

Its use does not demand a great level of competence, and it can be used by inexperienced farmers as well.

These fertilizers come in a variety of forms and are substantially less expensive than expensive man-made fertilizers.

Shri Bhushan Goel established Prabhat Fertilizer & Chemical Works in 1974, and it has been a pioneer in the field of manufacturing and exporting agricultural products in India.

In Karnal, Haryana, the company possesses the most modern and expansive production facilities.

This technology-driven company has grown into a multidimensional agri-biotech firm with a strong focus on organic fertilizers made with the most cutting-edge ingredients proven to improve plant health.

Types of Organic Fertilizers

Plant and vegetable wastes, animal matter and excreta, and mineral sources are used to make organic fertilizers.

Organic fertilizers have a complex biological structure, which is one of their main advantages.

They are always more easily available because they are made from locally sourced ingredients, and some of the organic fertilizers include:

- **Composts:** Organic waste that has decomposed by composting is referred to as compost.

Vegetable and plant waste, as well as animal excreta, are examples of organic materials.

- **Manure:** Manure is formed of faces from animals (cow dung & goat droppings).

Goat manure is heavy in nitrogen and potassium, while cattle dung is high in nitrogen and organic carbon.

- **Vermipost:** It is the result of numerous worm species degrading organic material, resulting in a diverse combination of decomposing food waste.

- **Bone Meal:** It's made up of animal bones and other ground slaughterhouse waste. Phosphorus and amino acids are abundant in this food. Because it's organic, it's also a slow-release fertilizer.

List of organic fertilizers

Prabhat Fertilizer & Chemical Works appears to specialize in two organic fertilizer types:

1- Prabhat Phosphate Rich Organic Manure

2- Prabhat Fertilizer Prabhat Shakti

Prabhat Phosphate Rich Organic Manure: Prabhat Phosphate Rich Organic Manure is processed using cutting-edge technology in the most optimal environment, ensuring that it is impurity-free, high in nutritional value, and has a long shelf life.

Prabhat PROM is significantly less expensive than chemical fertilizers, and it comes in a range of container sizes to suit the client's demands.

Prabhat Phosphate Rich Organic Manure is a balanced fertilizer that can be used in place of Diammonium Phosphate, Single Super Phosphate, and Rock Phosphate.

To encourage extensive plant growth, the water-soluble Prabhat PROM is mixed into the soil.

Advantages of phosphorus-rich organic manure include its ability to deliver phosphorus to the second crop planted in a treated region just as well as the first.

As well as its ability to be formed from acidic waste materials collected from biogas plant outflow

Pulses, grains, vegetables and fruits of all seasons are among the crops addressed.

Prabhat Shakti: Prabhat Shakti is a non-toxic, environmentally friendly organic manure that provides vital nutrients including nitrogen, phosphorus, and potassium in an organic form with a low carbon-to-nitrogen ratio.

It also contains beneficial microorganisms that are immobilized for better soil health.

To boost fertilizer use efficiency, micronutrients and secondary nutrients such as calcium and magnesium, as well as potassium as potash sulphate, have been added.

Prabhat Shakti enhances the fertility, texture, and organic matter content of the soil.

With a low C:N ratio, it delivers vital elements including nitrogen, phosphorus, and potassium in an organic form.

It also contains helpful microorganisms for better soil health when immobilized.

To boost fertilizer utilization efficiency, micronutrients and secondary nutrients such as calcium and magnesium, as well as potassium as potash sulphate, have been added.

This particular organic fertilizer is also compatible with Pulses, grains, vegetables and fruits of all seasons.

Benefits of Organic Fertilizers

Some of the guaranteed benefits that is assured if organic fertilizers are regularly used on plants-but in moderation are as follows;

- Soil texture is improved.
- Leaves of the plants are green and retain the actual color.
- Retains the nutrients and brings back the lost ones from the plants.
- Organic fertilizers unlike the synthetic fertilizers help in microbes thrive.
- Organic fertilizers are user friendly and human safe.
- Alongside, they assure environment safety by being economic and environment friendly.

Importance of organic fertilizer

Fertilizers can be used in a variety of ways in the garden or on the land. Organic fertilizer is no different.

We have created a list of the top ten benefits of utilizing organic fertilizer pellets.

Before you go on, you should know that Prabhat Fertilizer makes organic fertilizers utilizing a unique composting technology.

Animal faces is employed in the production of the products, which are then converted into organic fertilizer.

The fertilizer is subsequently compacted into pellets, resulting in a completely organic and biodegradable product.

The composting process provides some of the following benefits of organic fertilizers and these are as follows:

- **Made with 100% natural ingredients:** Our organic fertilizers are made with no residual flow additives. Considering our manufacturing process is built on a circular concept, our products are completely natural.

- **Organic fertilizer has a great capability for absorbing moisture:** Composting results in an organic fertilizer pellet that absorbs a lot of moisture (up to three times its own weight). In poor soil, moisture (rain) evaporates or is washed away quickly. As a result, less water will be wasted, and hence less water will be used.

- **It is long lasting:** Another benefit of organic fertilizer is that it may be stored easily and for extended periods of time.

- **The nutrients are slowly released and give ample amount of time for plants to heal:** Organic fertilizers release nutrients gradually, allowing the crop to absorb them while growing.

This ensures that the effect is both long-lasting and effective.

- **Organic fertilizers are applicable anytime and in all seasons of the year with ease:** Organic fertilizers can be applied to any soil type at any time of year (with the exception of frost in the ground).

During the growth season, the pellets can be disseminated manually or mechanically using standard technology. Mineral washout is decreased, and growth is accelerated as a result.

Prabhat Agri Organic Fertilizers encourages gardening to keep your soil, plants, and beneficial insects fertile.

This is accomplished by substituting organic matter and soil enrichment for synthetic fertilizers and insecticides.

Organic fertilizers can help you develop a bumper crop of vegetables or flowers, allowing your farm and garden to reach their full potential.

When you use organic fertilizer, your plants and vegetables will benefit from a balanced, nutrient-rich ecosystem that can function as nature intended.

Excess fertilizer, on the other hand, can result in nutrient loss, surface and groundwater contamination, soil acidification or basification, loss of beneficial microbial communities, and increased sensitivity to destructive insects, therefore use it wisely and according to the instructions.

As a result, we now know that organic fertilizers are far safer to use than synthetic fertilizers due to their lower concentration.

Organic Fertilizers since they are non-toxic and good for the environment, organic fertilizers are the best fertilizers for plants and crops in farming.

Their regular use has no negative impact on the environment and contributes to a brighter future.

Organic fertilizers help to eliminate unwanted and dangerous environmental contamination, such as surface water contamination, subsurface water table contamination, and so on.

They contribute to the degradation of other natural substances in the soil, enhancing its content.

Furthermore, because organic fertilizers have no negative side effects, fertilizing the soil too much will never harm the plants.

Biodegradable, renewable, long-lasting, and environmentally friendly, organic fertilizers are the way to go.

They ensure a steady supply of important nutrients and replenish the decreased vitamin and mineral content of the soil.

Nowadays, there are a plethora of high-quality organic fertilizers on the market. Prabhat Fertilizer produces some of India's best organic fertilizers.

Its high-quality products are used by farmers, landscapers, gardeners, and homeowners all over the world to help plants grow stronger.

Excessive and Disproportionate Use of Chemicals Cause Soil Contamination and Nutritional Stress

Soil is a very important and sensitive resource of a nation. In order to meet increasing public needs and to promote crop products, the use of high inputs of chemicals in the soil in the form of fertilisers, pesticides, fungicides, insecticides, nematicides and weedicides, along with intensive irrigation practises, helped to achieve the target to a certain stage. However, the decrease in crop yield took place despite the application of fertiliser.

The toxic chemicals influence the life of beneficial soil microorganisms, which are indeed responsible for maintaining soil fertility. Moreover, groundwater, air, and human and animal health have also been adversely affected by these chemicals directly and indirectly. Therefore, preserving the health of the soil is very essential. The avoidance of chemical fertilisers and the use of natural fertilisers such as biofertilizers, vermicompost, green manure and biopesticides, as well as the nourishing of the soil and the environment, can be a sustainable approach to crop productivity.

In order to boost crop quality and satisfy the global demand for food, chemical formulations being introduced as fertilisers and pesticides in appropriate amount are important for food management resources in agriculture. On the other hand, if used in excessive and disproportionate amount, there are harmful aspects of inorganic fertilisers and pesticides that can not be ignored.

They persist for a long time in the soil and atmosphere and influence various biotic and abiotic factors. They negatively influence soil, microflora, other organisms, human health and the environment. The excessive quantities of agrochemicals, industrial chemicals, trace metals and urban waste enter the soil through atmospheric deposition, disposal of waste, industrial effluents and direct application, and pollute it [125- 127].

Soil contamination is responsible for decreasing the soil biodiversity and fertility and hence, decrease soil health by obstructing the breakdown of soil organic matter and altering nutrient cycling. The contamination of soil, therefore reduces crop yield and affects food safety, especially when bioconcentrated pollutants enter organisms within food chains [128].

Through their roots, plants can also take up soil pollutants or absorb them through their leaves. The prolonged intake of infected foods, including human beings, can cause disease and lead to animal deaths [129]. In particular, urbanisation causes soil contamination in peri-urban areas, which have to deal with urban air pollution deposition and municipal solid waste disposal [125, 130].

Acid rain decreases soil aggregate stability, affects soil microorganisms and enzyme activities, increases soil erosion and mobility of nutrients, and in turn contribute to loss of nutrients [125, 131- 133].

Soil pollution is also linked to the water quality used for irrigation purposes and to flooding events. Industrial and urban sewage is rapidly being adopted for irrigation to meet the rising demand for drinking water. This is particularly common in countries facing rapid urbanisation rates, such as China, where contaminated water and sewage have irrigated 3.62 million ha of agricultural land [125].

Due to atmospheric accumulation, industrial waste and the use of waste water for irrigation, soil contamination from trace metals is prevalent in peri-urban areas [125, 134, 135]. Trace metal supply is directly influenced by plant root exudates and by the activities of soil microorganisms.

Owing to the high sensitivity of soil microorganisms to excessive trace metal concentrations, they are responsible for reducing soil biodiversity and fertility [136, 137].

Moreover, due to their close affinity with organic matter, trace metals accumulate in surface organic deposits, and passively taken up by plants by water flow [138]. Studies have documented the accumulation of trace metals in agricultural foods with high concentrations in stems and leaves rather than in fruits and seeds [139].

Intensive cultivation and mono-cropping have contributed to a large increase in the usage and reliance on agrochemicals. Fertiliser and pesticide soil contamination is due to inadequate management of nutrients in combination with pest and weed mismanagement, respectively [128].

In addition, because their behaviour in the environment and especially in the food chain is not comprehensively understood, the fate of pesticide metabolites is of concern [140]. The growth of the population raises the risk of pollution of the soil. Food protection is thus threatened by the presence of toxins and by the associated risks of bioaccumulation. Soil contamination is responsible for reducing crop yields and for turning productive agricultural land into unproductive areas by decreasing soil fertility and biodiversity.

As a result, the food availability and stability dimensions of food security are affected by this. Food accessibility is challenged by the extent and spatial distribution of soil pollution, which, in particular in urban and peri-urban areas, restricts food access. Soil pollution is, therefore, a hazard to all dimensions of food safety.

Healthy soils: a prerequisite for sustainable food security

Soil health is defined as a living soil's ability to function within natural or regulated ecosystem boundaries, to preserve productivity of plants and animals, to conserve and enhance the quality of water and air, and to promote plant and animal health [141, 142].

Therefore, soil health is a multi-dimensional and holistically vital soil characteristic, and forms the basis for healthy food production, thereby contributing to local and global food security. By 2050, a 60 percent rise in global food production and related ecosystem services need to be accomplished. But, through soil erosion, nutrient loss, salinity, sealing and pollution, one-third of global soils are currently facing moderate to extreme degradation.To achieve sustainable soil management, evidence-based decisions and soil information are essential [143].

Soils impoverished by nutrients lead to systemic food and nutritional problems. Micronutrient deficiencies are significant cause of morbidity and mortality, and affect over two billion people [144-147]. Protein-energy malnutrition is due to food scarcity and ingestion of trace elements (i.e. iron, lithium, magnesium, zinc, copper, iodine) from crops with low tissue concentrations, which are directly attributable to nutrient-poor soils [148].

Impact of chemicals used in agriculture on environment

Since the chemical fertiliser increases the growth and vigour of the plant, it thus meets the world's food security, but the plants grown in this way do not develop good plant characteristics such as good root system, shoot system, nutritional characteristics and will not have time to grow and mature properly [149].

The deleterious effect of chemical fertilisers itself begins with the processing of chemicals whose products and by-products are certain harmful chemicals or gases that cause air pollution, such as NH_4, CO_2, CH_4, etc. And it will cause water pollution when the waste from industries is disposed of untreated in nearby water bodies.

It also involves the most damaging impact of the accumulation of chemical waste in the bodies of water, i.e. water eutrophication. And its constant use, when applied to the soil, degrades the health and quality of the soil, thereby causing soil contamination. It is therefore high time to realise that our climate and biodiversity are depleted by this crop production input.

Therefore, its continuous use without taking any remedial action to reduce or judicious use will one day deplete all natural resources and threaten the entire life of the earth. The adverse effects of these synthetic chemicals on human health and the environment can be reduced or eliminated by adopting new agricultural technological practises, including the use of organic inputs such as manure, biofertilizers, biopesticides, slow-release fertilisers and nanofertilizers, etc., and moving away from chemical intensive cultivation.

Influence on soil compaction and degradation

Soil compaction is an important component of the syndrome of land degradation and is a major problem for advanced agriculture, affecting soil resources adversely [150].

As the soil is compacted, its composition changes by crushing aggregate units, decreasing the size of pore spaces between the soil particles, decreasing compaction due to the use of heavy equipment, reducing the use of organic fertiliser, repeated use of chemical fertilisers, and ploughing for several years at the same depth [151].

One of the principal causes of compaction is the usage of fertilisers more than the recommended amount for long periods and intensive cropping. Soil compaction causes problems such as excessive soil strength, root growth restriction, poor aeration, poor drainage, runoff, erosion and deterioration of the soil, etc. [152]. Such modifications lead to permeability, hydraulic conductivity and groundwater recharge reductions [153].

Excessive soil compaction impedes root growth and this decreases the capacity of plants to absorb nutrients and total porosity, leading to an increase in the density of soil bulk and resistance to penetration. It is reported that compaction decreases both root growth and yield by more than 80 percent [154]. Nitrification decreases by 50 percent as the density of soil bulk increases and plants consume less N, P and Zn from soil [155].

A great concern is the reduction of biological activities in soil due to compaction [156]. The most significant element in soil structure stability is organic matter. Soil that has high organic matter content and thrives with soil species is more compaction-resistant and can recover much better from mild damage to compaction [157, 158]. Over-use of fertilisers has led the development of continuous monoculture cropping, accumulation of fertiliser mineral salts in soil that forms compaction layers in soil, and cause long-term soil degradation.

Disproportionate usage of chemicals and soil nutrients

The soil is a home for soil organisms which are a mechanism for nutrient recovery, and offers many other environmental services. Chemical fertiliser overuse can contribute to soil acidification and soil crust, thereby reducing the content of organic matter, humus content, beneficial species, stunting plant growth, altering the pH of the soil, growing pests, and even leading to the release of greenhouse gases. The acidity of the soil reduces crop phosphate intake, raises the concentration of harmful ions in the soil and inhibits crop growth [35].

The soil's loss of humus decreases its capacity to store nutrients. The atmosphere is polluted by greenhouse emissions resulting from the excess use of nitrogen fertiliser. Over the time nitrogen fertilisers added in large quantities to fields kills the balance between the three macronutrients, N, P and K, resulting in decreased crop yields. Sandy soils are much more vulnerable to soil acidification than clay soils.

Clay soils have the potential to buffer excess chemical fertilisation effects. Repeated chemical fertiliser applications may lead to a toxic build-up in the soil of heavy metals such as arsenic, cadmium, and uranium. Not only do these toxic heavy metals pollute the ground, but they also accumulate in food grains, fruits and vegetables. Fertilisers such as triple superphosphate, have trace elements such as cadmium and arsenic that accumulate in plants and enter humans via food chains that can cause health problems [160].

Application of fertilisers without the recommendation of soil testing can lead to implications such as soil degradation, nutrient imbalance, soil structure destruction, bulk density increase [161].

When crop plants are harvested, soil nutrient levels are reduced over time, and these nutrients are replenished either by natural decomposition or by adding fertilisers. Therefore, the basic component of modern agriculture thesedays is fertiliser. However, while chemical fertilisers are the main cause of adequate crop production for the world's population, their overuse presents serious challenges for present and future generations, such as contaminated air, water and soil, degraded land, soils and increased greenhouse gas emissions.

Not only are these synthetic fertilisers being harmful to our climate, but also to humans, livestock, and microbial forms of life. It is high time that everyone realises the detrimental effects of using excess chemical fertilisers and take steps to minimise the usage of chemical fertilisers and pesticides by substituting other organic modifications such as organic manures that not only provide plants with essential nutrients, but also preserve soil quality for subsequent crops.

There are so many other technologies that are being developed, such as slow or controlled released fertilisers, prilled or granulated fertilisers, inhibitors of nitrification, nano-fertilisers, etc., all of which are the promising alternatives that can be used to solve these serious challenges and save both our environment and the ecosystem [159].

Microbial community structure

Soil microorganisms play an important role in the conservation of soil fertility and ecosystem work [162, 163]. The plant roots secrete carbon-containing organic material in the rhizosphere which is the source of carbon, nitrogen and energy needed for the growth and reproduction of soil microorganisms.

A large number of microbes gather around plant roots, which results in a distinction between the state of soil nutrients and the composition of the soil microbial population [164]. The region with the greatest contact between plant roots, soil and microorganisms is the rhizosphere. Microbes of the rhizosphere play an important role in the cycling of soil material and the transfer of energy.

Fertiliser application is an important management measure in agricultural production that not only promote crop growth and yield but negatively influence the soil microorganisms as well [165]. The widespread use of chemical fertilisers currently leads to a decline in soil fertility and a number of environmental problems, while bioorganic fertiliser not only improves soil fertility through the contribution of beneficial microorganisms and organic materials, but also eliminates many of the environmental problems caused by chemical fertilisers.

Studies have shown that various fertilisation treatments have a significant effect on the structure of soil microbial biomass and the community. Different applications of fertilisers change the physical and chemical properties of the soil, which in turn affects the structure of the soil bacterial community. Previous studies have found that pH, nitrate, and available phosphate and potassium are significant soil factors that influence the structure of the microbial community [133, 166].

By direct effects on the quality of soil nutrients, fertilisation affects soil microbial diversity. In conjunction with other mineral fertilisers, the long-term application of nitrogen fertiliser influences the nitrogen cycle and associated bacterial populations. Repeated overuse of chemical fertiliser may have a detrimental impact on the quality of soil and the composition of the soil microbial population. Long-term use of chemical fertilisers can dramatically decrease soil pH, which is closely related to reduced bacterial diversity and major changes in the composition of the bacterial population [1567].

Potential of biofertilizers to replace chemical fertilisers

As the land for agriculture is restricted and even diminished over time, the worldwide increase in the human population poses a major threat to the food security [168]. It is therefore important that agricultural productivity should be dramatically improved over the next few decades in order to meet the high demand for food from the emerging population. Furthermore, too much reliance for crop production on chemical fertilisers ultimately affects both environmental ecology and human health with great severity. A biofertilizer is a material that contains living microorganisms that colonise the rhizosphere or the interior of plants when applied to seeds, plants or soil and encourage plant growth by increasing the host plant's supply of nutrients [169].

The use of microbes as biofertilizers in the agricultural sector is considered an alternative to chemical fertilisers because of their wide potential to increase crop production and food safety [162]. Extensive work on biofertilizers has revealed their ability to supply the crop with the requisite nutrients in sufficient quantities to increase crop yield. Biofertilizers are widely used to accelerate certain microbial processes that increase the availability of nutrients that can be easily assimilated by plants. By fixing the atmospheric nitrogen and solubilising insoluble phosphates, biofertilizers increase soil fertility and produce plant growth-promoting substances in the soil [169].

The naturally accessible biological system of nutrient mobilisation, which greatly increases soil fertility and ultimately crop yield, has been encouraged by biofertilizers.

Biofertilizers are expected to be a healthy alternative to chemical inputs and to a great extent mitigate ecological disruption. Biofertilizers are cost-effective in nature, eco-friendly, and their extended usage greatly increases soil fertility. It has been stated that the use of biofertilizers increase the protein content, essential amino acids, vitamins, and nitrogen fixation, thereby increases crop yield by about 10–40 percent [170].

The advantages of using biofertilizers include low-cost nutrient sources, excellent microchemical and micronutrient suppliers, organic matter suppliers, growth hormone secretion, and the counteraction of chemical fertiliser adverse effects. Microbes are important soil components and play a crucial role in the different biotic activities of the soil ecosystem that make the soil dynamic for the mobilisation of nutrients and sustainable for the production of crops [171].

Improving soil fertility

Physical fertility refers to the soil's physical properties, its composition, texture, water holding properties, the way water flows to the roots of plants, and how the soil is penetrated by those roots. Biological fertility refers to the species and their capacity to play important roles that live in the soil. A soil's composition, its acidity or alkalinity, and its ambient temperature are only a sample of the several variables that decide the degree to which plants have access to nutrients [133].

The relative value of these variables depends on the nutrients, the soil and the plant. Most notably, soil structure determines how well the soil holds nutrients and water. Organic matter-containing clays and soils retain nutrients and water much better than sandy soils. The microbial community of the soil would also be highly influenced by the soil structure. If the soil does not allow these species to survive, plants that rely on bacteria or fungal species for nutrient uptake will not grow. Until recently, the application of fertiliser was the most commonly used treatment for nutrient deficiency.

As plant nutrient requirements vary over the plant life cycle, timing is also important. The effect of Liebig 's Law may obscure the identification of genuinely deficient nutrients, as the correct scarce nutrient may not be directly recognised by deficiency symptoms. Additional fertiliser would be of little to no assistance if the requisite structural and biological conditions are not present. The improved vitality of plants would rely on improving the structural and biological fertility of the soil.

Inappropriate application of fertiliser is a waste of time and money, but it can also have dire environmental implications as well [143].

Healthy soil is that which allows plants to grow to their maximum productivity without disease or **pests** and without a need for off-farm supplements. Healthy soil is teeming with bacteria, fungi, algae, protozoa, nematodes, and other tiny creatures. Those organisms play an important role in plant health.

Soil bacteria produce natural antibiotics that help plants resist disease. Fungi assist plants in absorbing water and nutrients. Together, these bacteria and fungi are known as "organic matter." The more organic matter in a sample of soil, the healthier that soil is.

Research conducted at Rodale Institute has shown that organic systems increase soil organic matter and soil health over time

To determine if soil is healthy, farmers and scientists measure several factors. How many microorganisms are present? How many nutrients—nitrogen, for example—are in the soil? How well does the soil retain water during drought? How much carbon can the soil **sequester** from the atmosphere?

At Rodale Institute, our scientists collect soil samples in the field. Back in the lab, they dry and weigh the samples before analysis. Our research has shown that while conventional systems erode and deplete soils, organic systems **improve and build** the soil over time.

Soil samples dry in the lab

Healthy soil contains aggregates that help it bind together, **preventing erosion and run-off**. It can **hold more water**, so plants fare better in drought. It contains more bacteria and fungi that help plants **fight diseases and pests**. And healthy soil also contains **more minerals and nutrients** that feed plants.

Healthy soil is the foundation of our global food system, but currently, it's at risk. The United Nations reports that using current practices, we have fewer than **60 years** of farmable topsoil remaining.

Every **organic farming practice** contributes to healthy, resilient soil that can support abundant life both below and above ground, making organic farming a powerful tool for soil conservation.

Healthy soils, a necessity for the EU

Maintaining and restoring our soils is key to achieve climate neutrality, zero pollution, sustainable food provision and a resilient environment. However, they are a fragile and finite resource and they are currently at risk, in Europe and beyond. This is why healthy soils are at the core of the European Green Deal.

In a recent article, scientists and policymakers from the Commission outline the key policies to achieve the ambitious soil objectives of the EU Green Deal.

Moreover, today the European Innovation Partnership for Agricultural Productivity and Sustainability is organising a brokerage event for farmers, researchers, and other relevant stakeholders to get involved in 'the EU Mission: A Soil Deal for Europe'.

Soil hosts more than 25% of all biodiversity and provides 95-99% of food to 8 billion people. It provides us with clean water and it is the largest terrestrial pool of carbon. Keeping our soils healthy is essential for life on Earth.

The EU Soil Mission has recently opened a call for €95 million to identify strategies for land decontamination and re-use, and innovative ideas on how to implement carbon-farming practices, e.g. methods to maintain and enhance carbon in soils.

Soil degradation – a serious and growing problem

Soil degradation in the EU is an increasing problem that threatens our wellbeing, our economy and our prosperity.

In fact, scientists from the Joint Research Centre estimate that 13% of soils in the EU suffer from high levels of erosion with annual costs of € 1.25 billion due to loss in agricultural productivity. Every year, cultivated lands lose 7.4 million tonnes of carbon due to non-sustainable management. And according to a European Court of Auditors report, 25% of land in southern and eastern Europe is at high risk of desertification.

The problem is further exacerbated by the loss of soil resources due to covering the land for housing, roads or industry ('soil sealing') - predominantly at the expense of agricultural land-. The annual net land take was estimated at 440 km²/year in the period 2012-2018.

In addition, contamination with harmful substances damages soil quality. The JRC reported around 2.8 million potentially contaminated sites in the EU and neighbouring countries.

In 2020, the European Commission presented an ambitious package of measures consisting of the biodiversity strategy for 2030, the Farm to Fork Strategy and the chemicals strategy as well as the circular economy action plan and the European Climate Law, all of which included actions to protect soils.

These were complemented by others, such as the Fit for 55 package, the zero pollution action plan, climate adaptation strategy, EU forest strategy, organic action plan, long term vision for rural areas and the EU soil strategy for 2030. In addition, the European Commission will propose a Soil Health Law in 2023.

To support the above-mentioned EU policy initiatives, an EU Soil Observatory (EUSO) has recently been established. It will also help to monitor agricultural soils in the context of the Common Agricultural Policy (CAP) and the Clean Soil Outlook of the Zero Pollution Action Plan.

Specifically, the EU Soil Observatory aims to

1.   monitor soil-related EU policies with an enhanced EU Soil Indicator Dashboard,

2.   serve as an open forum for increasing awareness,

3.   provide a stronger European Soil Data Centre,

4.   integrate national soil monitoring systems with Land Use/Cover Area frame Survey (LUCAS) and

5.   support to Horizon Europe Research and Innovation programmes.

© EU 2022

European Commission's Soil Mission

In September 2021, the European Commission adopted five Research and Innovation Missions to bring concrete solutions in response to major societal challenges.

The Mission "A Soil Deal for Europe" will support the EU's ambition to manage land in more sustainable ways, thereby supporting the EU's Green Deal agenda and meeting global commitments such as the Sustainable Development Goals (SDGs).

The Soil Mission will establish 100 living labs and lighthouses to co-create, test and pioneer innovations for soil health at local level. It will also help develop a harmonised framework for soil monitoring in Europe, increase soil literacy in society and raise awareness on the vital importance of soils.

The EUSO will both support the implementation of the Soil Mission, and benefit from the outcomes of the Mission's research and innovation on soils. The observatory will also focus on raising societal awareness of the value and importance of soils to the lives of citizens.

One of the main objectives of the mission is to build a robust, harmonised EU framework for soil monitoring and reporting.

Soil health is defined as the continued capacity of soil to function as a vital living ecosystem that sustains plants, animals, and humans. Healthy soil gives us clean air and water, bountiful crops and forests, productive grazing lands, diverse wildlife, and beautiful landscapes. Soil does all this by performing five essential functions:

- **Regulating water**

  Soil helps control where rain, snowmelt, and irrigation water goes. Water flows over the land or into and through the soil.

- **Sustaining plant and animal life**

  The diversity and productivity of living things depends on soil.

- **Filtering and buffering potential pollutants**

  The minerals and microbes in soil are responsible for filtering, buffering, degrading, immobilizing, and detoxifying organic and inorganic materials, including industrial and municipal by-products and atmospheric deposits.

- **Cycling nutrients**

  Carbon, nitrogen, phosphorus, and many other nutrients are stored, transformed, and cycled in the soil.

- **Providing physical stability and support**

  Soil structure provides a medium for plant roots. Soils also provide support for human structures and protection for archeological treasures.

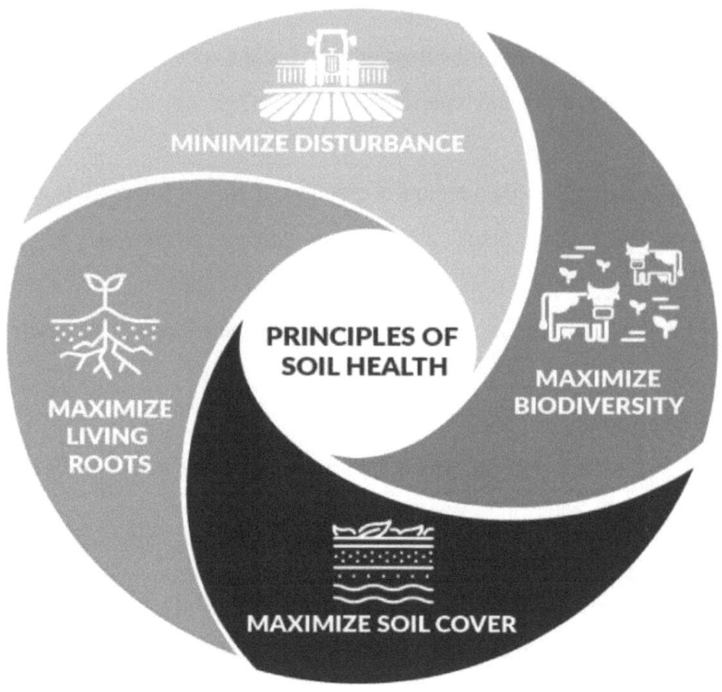

Principles to Manage Soil for Health

Soil health research has determined how to manage soil in a way that improves soil function.

- **Maximize Presence of Living Roots**
- **Minimize Disturbance**
- **Maximize Soil Cover**

- **Maximize Biodiversity**

As world population and food production demands rise, keeping our soil healthy and productive is of paramount importance. By farming using soil health principles and systems that include no-till, cover cropping, and diverse rotations, more and more farmers are increasing their soil's organic matter and improving microbial activity. As a result, farmers are sequestering more carbon, increasing water infiltration, improving wildlife and pollinator habitat—all while harvesting better profits and often better yields.

The Components of Healthy Soil

"Soil is the source for all food and fiber consumed on Earth. It is a medium composed of minerals, organic matter, water, air and living organisms. It provides infrastructure to support and nurture plants." Let's take a brief look at each of these key components.

1. Minerals

There are numerous different minerals, all charged with an important task to promote plant and soil health. Some of the key minerals include:

Carbon – The building block of all organic chemistry.

Calcium – In plants, calcium is a structural component of the cell wall. In plants, fungi and bacteria, calcium is a cofactor in regulating internal chemical processes. In soil, calcium shifts the pH to alkaline (vs acid).

Manganese – An essential element of chlorophyll (think Photosynthesis)

Nitrogen – All plant and animal proteins contain nitrogen. It is essential to life.

Potassium – An essential plant nutrient involved in the regulation of photosynthesis and water management in plants.

Phosphorous - an essential element in the biochemistry of photosynthesis, metabolism of sugars, energy management, cell division and the transfer of genetic material. In fact, phosphorous is essential to all life forms.

Sulfur – An important component of amino acids and proteins.

2. Organic Matter

3 to 5% of soil consists of organic matter. It is derived from dead and decaying plants and animals. It contributes to soil nitrogen, phosphorous and sulfur. The end result of organic matter is humus, which is created after microorganisms break down the organic matter.

3. Water

Water acts as a solvent and carrier for plant nutrients. Just like all animals, microorganisms require water for their metabolic processes.

4. Soil Air/Atmosphere

50% of soil volume consists of pore spaces filled with air and/or water. Gases found in soil air include: oxygen, nitrogen, carbon dioxide and water vapor. However, due to respiration of soil microorganisms, soil air has more carbon dioxide than the air we breathe (the atmosphere above the soil).

5. Soil Microorganisms

There are more microbes in 1/3 cup of soil, than there are people on earth! Soil microorganisms include: bacteria, fungi, algae, protozoa, arthropods, nematodes and worms. Often, tilled soil (soil used in agriculture and landscaping) is more bacterially dominant while non-tilled soil (soil found in a forest) is more fugally dominant.

Environmental Performance in the Production and Use of Recovered Fertilizers

The linear economy model based on the use of fossil fuel and raw sources has led our planet to encounter major environmental problems such as climate change, land degradation, and alteration of biochemical cycles. [172] With particular reference to N and P global flows, it has been reported that the current uses of these two elements are over Earth's boundaries because of anthropogenic perturbation due, mainly, to fertilizer application. [173]

The use of chemically produced N and mined P is modifying and misbalancing not only the agroecosystem but also the natural ecosystems, putting biodiversity at risk. [174]

The regular production and use of mineral fertilizers in agriculture have a long track record of impacts on the environment beyond the mere addition of nutrients to the soil.

Fertilizer industry production and use causes about 2.5% (1203 Tg CO_2 equiv) of global GHG emissions, [175] and N fertilizers account for 33% of the total annual creation of reactive N, i.e., 170 Tg N y^{-1} (fertilizers and livestock manure), [176, 177] generating big environmental problems. In addition, the production of P and K fertilizers relies upon nonrenewable and extracted resources that are becoming depleted[178] and are concentrated (e.g., P) in only a few countries. [179]The consequence of that is the need for new management strategies to reduce the additions of N and P into the ecosystem with particular reference to agriculture.

The Circular Economy has been indicated as a new productive paradigm to produce goods, and it consists in the redesign of productive processes to allow the successive recovering of wastes for new productive processes, avoiding the use of new resources. [180]

Organic wastes can be explored as raw materials to recover nutrients and organic matter, representing an example of Circular Economy. To do so, wastes should be accurately chosen so that nutrient recovery can be made by applying suitable technologies, [181] producing fertilizers to replace synthetic ones. [182]

Anaerobic digestion (AD) is a suitable biotechnology for producing biofertilizers, thanks to the process that modifies organic matter and the nutrients it contains, resulting in a good amendment and fertilizer properties of the end product, i.e., digestate. [183-185] In addition, the AD process renders the digestate more suitable for subsequent biological/physical/chemical treatments allowing organic matter (OM) and N and P to be separated, producing both an organic amendment and N and P fertilizers. [181, 186- 188]

The recovery of nutrients allows the production of fertilizers able to substitute for synthetic ones, thus reducing the necessity to produce fertilizers using fossil energy (N and P) and fossil resources (P and K), [189] and closing nutrient cycles. In addition, the recovery, also, of the organic matter represents a solution to the problem of low organic matter (OM) content (<1%) of soils, [190] which are attributed to the high carbon dioxide emissions which result from the intensification of agricultural practices. [191]

Despite the clear need to better manage nutrients already present in the ecosystem without adding new ones, a significant obstacle to this is the low efficiency and environmental performance, which have been attributed to recovered nutrients. [176,192]

Synthetic fertilizers contain concentrated nutrients under available forms, and so they are easy to apply to meet crop requirements. By contrast, the recovered wastes (sewage, manure, digestates, etc.) contain nutrients with low efficiency and low concentration, and which also require good practices to be used to avoid environmental impacts. [193, 194]

Low-nutrient use efficiency (NUE) of recovered fertilizers might be due to their nonappropriate chemical form (mineral vs organic forms), loss as NH_3 volatilization (10–65%), NO_3^- leaching and runoff (1–20%), and nitrification–denitrification (1–30%). [195, 196]

Therefore, the increase of NUE and environmental outcomes of recovered fertilizers represent challenges for modern agriculture. [197]

Recently, a scientific paper described, [181] at full scale, an AD plant producing recovered fertilizers (renewable fertilizers—RF) by anaerobic digestion, proposing that these fertilizers be used to substitute completely for fertilization by synthetic mineral fertilizers (SFs). Despite this, producing recovered fertilizers does not mean that they are capable of replacing synthetic ones ensuring better environmental performance. [198]

Therefore, the assessment of environmental impacts of recovered fertilizers needs to be studied in comparison with synthetic ones using an appropriate approach, i.e., life cycle assessment (LCA) fed by validated data (full-field data). To do this, the entire supply chain, i.e., production of fertilizers and their use in place of synthetic ones, must be considered.

The literature over the past 10 years has focused on renewable energy production through anaerobic digestion and less on the analysis of impacts related to the recycling of nutrients and their use. For example, Hijazi et al. [199] reviewed 15 LCAs related to anaerobic digestion of different biomasses focusing on the amount of renewable energy produced, used as a functional unit, rather than on recycling of nutrients, similar to other studies. [200, 201]

On the contrary, Timonen et al. [202] underlined the importance of AD in recovering nutrients, and LCA performed considered digestate production (renewable fertilizer) together with energy.

The authors compared three different digestate management to consider a multifunctional approach (bioenergy and fertilizers) capable of increasing the efficiency of the agricultural system. The work done underlined that the emissions due to the storage of digestate, its transport, and use in the open field were higher than those due to the use of chemical fertilizers.

Despite this, by combining the emissions due to anaerobic digestion and those due to the use of digestate, emissions were lower than those due to the production and use of mineral fertilizer. In the work cited, similarly to others, [198,201,203] data used (e.g., ammonia and N_2O emissions and nitrate leaching) were calculated using IPCC coefficients and no experimental data were used. Lyng et al. [204] underlined the importance of direct measurement from full-scale realities to avoid over- or underestimates.

This work likes to contribute to the existing literature by providing novelty in terms of the LCA approach able to consider the whole chain in producing and using recovered fertilizers from sewage sludge by AD. In particular, this work aims to complete the path of the proposed Circular Economy in agriculture, by measuring directly in the open field, the impacts derived from the use of recovered fertilizers, used according to virtuous approaches capable of reducing the resulting impacts. The full-scale approach and the use of directly measured data aim to correctly and experimentally evaluate the effectiveness and sustainability in the recycling of nutrients to replace synthetic mineral fertilizers.

Functional Unit

Functional Unit

The functional unit (FU) provided a reference to which all data in the assessment were normalized. Because this study considered the impacts derived from the production and use of fertilizers on maize crop, the functional unit chosen was referred to the fertilization (fertilizers production and use) of 1 ha of maize, i.e., for the Scenario SF: 402 kg of urea (185 kg of N), 476 kg of chemical ammonium sulfate (100 kg N), 195 kg of triple phosphate (89 kg of P_2O_5), and 165 kg of potassium sulfate (82.5 kg of K_2O), and for Scenario RF: 48 Mg of digestate, i.e., 370 kg of total N, i.e.,185 kg of effective N, 317 kg of P_2O_5, and 43 kg of K_2O, 1.38 Mg of recovered ammonium sulfate (100 kg of N), and 80 kg of potassium sulfate (40 kg of K_2O) (Table 3).

Table 3. Inventory Data of the Considered Scenario

	unit	quantity	data source
input			
waste input (total)	$Mg\ y^{-1}$	81 886	provided by facilitya
methane (from national grid)	$sm^3\ y^{-1}$	228 177	provided by facility
water (from aqueduct)	$m^3\ y^{-1}$	19 744	provided by facility
water (from well)	$m^3\ y^{-1}$	14 044	provided by facility
water (total)	$m^3\ y^{-1}$	33 788	provided by facility
electricity consumed from the grid	$kWh\ y^{-1}$	7189	provided by facility
sulfur acid	$Mg\ y^{-1}$	316	provided by facility
output			
digestate produced	$Mg\ y^{-1}$	112 322	provided by facility
electricity produced and fed to the grid	$kWh\ y^{-1}$	5 349 468	provided by facility
electricity produced and reused in the	$kWh\ y^{-1}$	2 395 215	provided by facility

	unit	quantity	data source
process			
total electricity produced	kWh y^{-1}	7 737 494	provided by facility
ammonium sulfate	Mg y^{-1}	571	provided by facility
wastes from sieving sent to landfill	Mg y^{-1}	2.5	provided by facility
biogas produced	Mg y^{-1}	3842	provided by facility
thermal energy produced (by CHP)	MWh$_{th}$ y^{-1}	5976	provided by facility
emissions (from distribution) digestate			
ammonia (N-NH$_4$)	kg ha^{-1}	25.2	detected on-site by the authorsb (Table S4)
direct dinitrogen monoxide (N-N$_2$O)	kg ha^{-1}	9c	detected on-site by the authors (Table S4)
indirect dinitrogen monoxide (N-N$_2$O)	kg ha^{-1}	0.8	IPCC 2006
nitrate leaching (N-NO$_3$)	kg ha^{-1}	83d	IPCC 2006
NO$_x$ (N-NO$_x$)	kg ha^{-1}	0.5	IPCC 2006
P surface run off (P)	kg ha^{-1}	1.4	EDIP 2003
urea			
ammonia (N-NH$_4$)	kg ha^{-1}	25.2	detected on-site by the authors (Table S4)
direct dinitrogen monoxide (N-N$_2$O)	kg ha^{-1}	9b	detected on-site by the authors (Table S4)
indirect dinitrogen monoxide (N-N$_2$O)	kg ha^{-1}	0.8	IPCC 2006
nitrate leaching (N-NO$_3$)	kg ha^{-1}	83c	IPCC 2006

	unit	quantity	data source
NO$_x$ (N-NO$_x$)	kg ha^{-1}	0.3	IPCC 2006
carbon dioxide (C-CO$_2$)	kg ha^{-1}	80.2	IPCC 2006
P surface run off (P)	kg ha^{-1}	0.2	Nemecek and Kägi 2007
use of nutrients			
RFe			
digestate	Mg ha^{-1}	48	data from authorsf
TN supplied by digestate	kg ha^{-1}	370	data from authors
TN delivered by ammonium sulfate	kg ha^{-1}	100	data from authors
P supplied by digestate	kg ha^{-1}	138	data from authors
K supplied by digestate	kg ha^{-1}	36	data from authors
K delivered as potassium sulfate	kg ha^{-1}	34	data from authors
SFc	kg ha^{-1}		
TN supplied by urea	kg ha^{-1}	185	data from authors
TN delivered by ammonium sulfate	kg ha^{-1}	100	data from authors
P provided by triple phosphate	kg ha^{-1}	39	data from authors
K supplied as potassium sulfate	kg ha^{-1}	70	data from authors

[a] Provided by facility: data acquired directly from the full-scale plant under study.

[b] Detected on-site by the authors: data acquired from open-field experimentation (see also the SI and Table S4).

[c] N$_2$O emissions were considered similar (calculated on 1 ha surface) for the two Scenarios as revealed by full-field measurements made after digestate and urea distribution (see Table S4).

[d] N leaching was assumed similar (calculated on 1 ha surface) for the two Scenarios as revealed by soil sampling made at 1 m soil depth in full-field trials (Table S4).

[e] RF: recovered fertilizer Scenario and SF: synthetic fertilizer Scenario.

[f] Data from authors: data derived from fertilization plan and fertilizer properties (Tables S1–S3).

Anaerobic Digestion Plant

The AD plant (1 MWe power) for the combined production of fertilizers and energy is situated in the Lombardy Region (North Italy). [181] The plant exploits anaerobic digestion (AD) to transform different organic wastes (sewage sludges produced by municipal WWTP, agri-food factories, and liquid pulp-fraction of source-separated domestic food wastes) into organic-mineral fertilizers, i.e., digestate, mineral N-fertilizer (i.e., ammonium sulfate), and energy (thermal and electrical). The plant is composed by two main sections comprising the AD plant and the ammonia-stripping unit (Figure 8a).

Figure 8

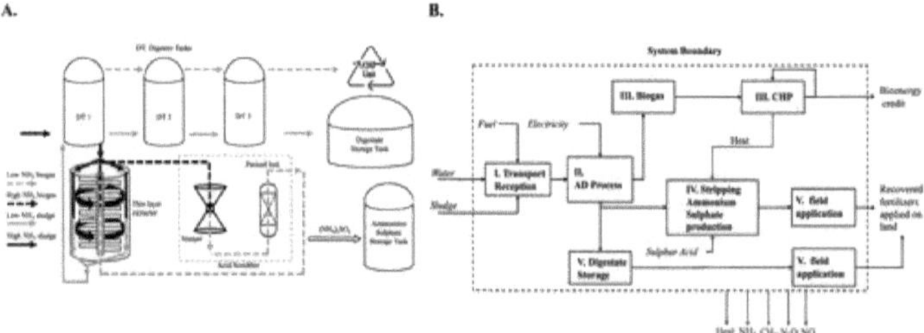

Figure 8. Anaerobic digestion (AD) plant and nitrogen-stripping unit layouts (a); system boundaries and main processes for the recovered fertilizers (RFs) (b).

The AD plant produces biogas that is exploited to produce electrical energy delivered to the national grid and is also used for plant autoconsumption and heat that is used for digester heating by steam injection and in the ammonia-stripping unit. During the process, several data were continuously monitored: digestate, pH (daily), digestate temperature, produced biogas, and biogas composition (CH_4, CO_2, and H_2S, this latter four measurement per day).

Anaerobic digestion takes place in three reactors, working in series, of 4500 m^3 each, made in carbon steel, with an average hydraulic retention time (HRT) of 45–50 days, which is longer than usual HRT for AD plant-treating sewage sludge but useful to ensure good biological stability and sanitation. [181]

The AD process is performed in thermophilic conditions (55 °C), where the temperature is kept stable using the heat produced from the combined heat and power (CHP) unit. Reactor tanks have no mechanical mobile parts inside, with digestate mixing guaranteed by a system of external pumps. The tanks are covered with a gasometric dome membrane and maintained at constant pressure.

The system withdraws digestate from the second digester tank (DT 2) (Figure 8a) to the thin layer extractor, where ammonia is stripped from digestate using the biogas or air. [181,206] The thin layer extractor consists of a cylindrical tank having inside a rotor with radial paddles, which by rotating at high speed keeps the digestate spread in a thin layer (few millimeters thick) on the internal walls of the cylinder.

Meanwhile, the rotor keeps biogas at high turbulence to enhance the exchange of ammonia from the digestate to the gas. The transfer of ammonia occurs in a counter current; the digestate is pumped into the top of the cylinder, and it goes down by gravity in a thin layer while gas flux is from the bottom to the top. The walls of the cylinder are warmed at 80 °C to increase the exchange from the digestate to the gas which is injected at 70 °C. After the stripping in the thin layer, the low-content ammonia digestate is pumped back to the first digester (DT 1) while carrier gas in a closed-loop cycle goes to the acid scrubber unit, where ammonia reacts with sulfuric acid-generating ammonium sulfate. Both recovered fertilizers produced were used in substitution for synthetic fertilizers, both at presowing (digestate) and as top-dressing (ammonium sulfate).

Recovered Fertilizers Produced

Recovered fertilizers (renewable fertilizers) characteristics are listed in Tables S1 and S2; a complete description can be found in Pigoli et al. (10) The previous characterization made also included organic contaminants and target-emerging organic contaminants (Table S1).

Full-Field Agronomic Use of Renewable Fertilizers in Substitution of Synthetic Mineral Fertilizers

Full-field agronomic performance and impact measurements, i.e., air emissions (NH_3, N_2O, CH_4, and CO_2) and nitrate leaching, were carried out on soil plots distributed randomly close to the AD plant. Digestate was injected into the soil at a depth of 15 cm at the dose required assuming a N efficiency of 0.5, as suggested by the Regional Plan for Water Protection from Nitrate from Agriculture. [207]

For the SF Scenario, urea was spread onto the soil surface following a routine agricultural procedure. The dosage of fertilizers was made according to common practices. Fertilizers used, doses applied, and spreading methodology are reported in detail in Table S3 in the Supporting Information and summarized in Table 3.

GHG emissions (N_2O, CH_4, and CO_2) were measured in 2020, following the entire agronomic season of maize: from May (sowing) to October (harvest). The determination of emissions was conducted through the use of non-steady-state chambers.[208]

Sampling chambers were placed in each of the experimental plots; furthermore, to obtain a background measurement, another three chambers were placed on nonfertilized plots. The air sampling inside the chamber was carried out with a frequency of 1–8 times a month, depending on the season and the state of the crop. The air taken was then analyzed in the laboratory using a gas chromatograph, according to the method reported by Piccini and colleagues. [209]

The cumulative emissions were obtained by estimating the flows in the nonsampling days, by linear interpolation. [210] The concentration of NH_3 was monitored by the exposure of α passive samplers. [193,211] For each plot, α samplers were installed in sets of three. To obtain background environmental concentration values, an additional sampling point was placed at a distance of about 1000 m away from fertilized fields and other possible point sources of NH_3 emissions.

System Boundaries and Data Inventory

System Boundaries

The system boundary starts from the organic waste collection and transport encompasses the production of digestate/biofertilizer and ammonia sulfate, the correlated processes for producing biogas which is transformed into electric energy and thermal energy and finally the use of the digestate in the field.

The system boundary is represented by the dashed line in Figure 8b and comprises five main processes for Scenario RF (recovered fertilizer): (i) the transport of sludge and organic wastes to the AD plant (assuming 100 km on average), (ii) the AD process, (iii) the biogas combustion and electricity production in CHP, (iv) the digestate stripping process and ammonium sulfate production, and (v) the digestate storage, handling, and distribution into fields. Capital goods were included in the system, considering a lifespan of the structure of 20 years.

The Scenario SF (synthetic fertilizer) encompassed the production of urea, triple phosphate, and potassium sulfate fertilizers (including logistics and transportation) and the timely distribution on fields. This Scenario was modeled using data coming from the literature and databases (Ecoinvent 3.6).[212]

The main data inventory is reported in Table 3; inputs and output of production were all taken directly from the plant facility. Air emission of the two systems, i.e., ammonia, methane, nitrous oxide, and carbon dioxide, was measured directly on monitored field plots as previously reported (Tables 3 and S4).

Indirect dinitrogen monoxide and NO_x were estimated according to IPCC. [213] Nitrate leaching was calculated according to IPCC [213] for the Scenario SF, based on the N distributed, and assumed to be equal for Scenario RF, as the monitoring of nitrate content in deep soil layers during the year showed no differences (Table S4). Phosphorus in soil, leaching, and run off was modeled according to Ecoinvent report 15. [214]

Heavy metals supplied were included in the model according to the characterization data of digestate, plant uptake, and accumulation rate in the soil system. [215,216] The input of organic pollutants was considered for PCDD/F, DEHP, and PAH contained in digestate, as a proper numerical quantification was workable (Table S1).

Modeling Framework and Approach to Multifunctionality

The modeling framework of this study was attributional, i.e., digestate and ammonium sulfate were considered as the target products of the production chain. Biogas was produced and valorized in the CHP module to generate electricity and heat.

To consider these outputs and to make the two systems (Scenario RF and Scenario SF) comparable, the approach of system substitution, i.e., crediting for the avoided burden, was chosen. The option of system substitution was not exploited to include the service of waste treatment (i.e., incineration or landfill) that is performed, as it would have introduced great variability in the credits of the service.

This approach was very prudential, as it did not consider the alternatives for disposal of organic wastes that in any case would be necessary and impacting. However, the credits for renewable electricity were accounted for and considered for substituting the electricity mix distributed in the national grid.

Life Cycle Impact Assessment

The life cycle impact assessment (LCIA) was based on the emissions and resource inputs identified during the data inventory, which was processed into indicators that reflect resource shortage and environmental burdens.

The software SimaPro Analyst 9.1.1.7 [217] was used for the computational implementation of the inventories and the set of libraries covered by Ecoinvent databases v3.6, 2019 to analyze environmental impacts. Because of its representativeness at the global scale, the ReCiPe 2016 method (version 1.13), [218] which contains midpoint impact indicators and end point areas of protection, was used to assess the environmental performance of biofertilizers and energy production. Global normalization factors from the same method were used. [219]

The robustness of the LCA results was assessed by Monte Carlo analysis, setting 10 000 runs. [220]

The two Scenarios reported as midpoint indicators and split for fertilizers production and use, as well as the impact deviations taking as reference Scenario RF, are shown in Table 4. The Scenario RF showed better environmental performances than the system encompassing the production and use of urea and commercial fertilizers (Scenario SF).

In particular, for the Scenario RF, 11 of the 18 categories showed a lower impact than in the Scenario SF, and four of the categories (ionizing radiation, terrestrial ecotoxicity, fossil resource scarcity, and water consumption) showed net negative impacts in the Scenario RF, getting the benefits from the credit of renewable energy production by AD. The final end point single score ranked 48 and 215 points for the Scenario RF and Scenario SF, respectively, which summarizes the globally better outcome of the Scenario RF (Figure 9). Analysis and contributions of the processes to the categories are discussed below.

Figure 9

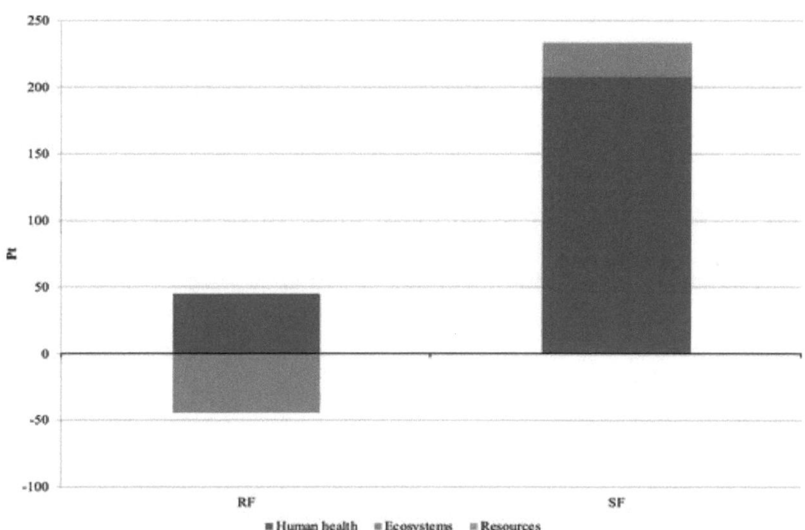

Figure 9. Comparative environmental results for Scenarios Recovered Fertilizers (RFs) and Synthetic Fertilizers (SFs). Impact assessment (Ecopoint—Pt) calculated according to the ReCiPe 2016 end point (H) V 1.03 impact assessment method.

Table 4. Impact Category Values for the Two Compared Systems SF and RF with Their Respective Contribution Due Production and Use (Field Emission and Distribution), and Credit Related for the Electricity Generated (CRE)a

		RF				SF		
impact category	unit	production	use	CRE	total	production	use	total
global warming	kg CO_2 equiv	669	3999	– 1315	3354	834	3966	4800
stratospheric ozone	kg CFC11	0	0.1	0	0.1	0	0.1	0.1

impact category	unit	production	use	CRE	total	production	use	total
		RF				SF		
depletion	equiv							
ionizing radiation	kBq Co-60 equiv	38	10	−204	−156	82	4.5	86
ozone formation, human health	kg NO$_x$ equiv	5	2	−3	4	1	1.0	2
fine particulate matter formation	kg PM2.5 equiv	2	6	−2	7	1	6.2	8
ozone formation, terrestrial ecosystems	kg NO$_x$ equiv	5	2	−3	4	1	1.0	2
terrestrial acidification	kg SO$_2$ equiv	6	50	−5	51	4	50	54
freshwater eutrophication	kg P equiv	0.1	8.4	−0.3	8.2	0.3	0.2	0.5
marine eutrophication	kg N equiv	0	17	0	17	0.0	17	17
terrestrial ecotoxicity	kg 1,4-DCB	1247	240	−1370	117	2550	114.8	2664
freshwater ecotoxicity	kg 1,4-DCB	8	351	−11	348	13	0.6	14
marine ecotoxicity	kg 1,4-	12	492	−16	488	23	0.9	24

impact category	unit	RF				SF		
		production	use	CRE	total	production	use	total
	DCB							
human carcinogenic toxicity	kg 1,4-DCB	35	9	–25	19	19	1.4	20
human noncarcinogenic toxicity	kg 1,4-DCB	266	54 585	–330	54 521	458	88.8	547
land use	m^2 a crop equiv	7	3	–4	6	6	1.1	7
mineral resource scarcity	kg Cu equiv	3	1	–1	4	9	0.4	9
fossil resource scarcity	kg oil equiv	134	27	–384	–224	313	16	329
water consumption	m^3	631	189	–8575	–7755	1196	86	1282

[a] Impact assessment calculated according to ReCiPe 2016 Midpoint (H) V.1.1. FU: 1 ha Maize.

Midpoint Results of Impact Categories Related to Ecosystem Quality

Global Warming Impact Category

The production of the recovered fertilizers (Scenario RF), which included sludge transport and handling, the AD process, ammonia stripping, and biogas burning, without considering the electricity credits, caused the emission of 669 $kgCO_{2equiv}$, lower than the data reported for the production of synthetic mineral fertilizers, i.e., 834 $kgCO_{2equiv}$.

Beyond, thanks to the credits (avoided CO_2 emissions) due to the production of renewable energy (biogas), the value of the fertilizers production was negative, i.e., -646 $kgCO_{2equiv}$. With reference to the fertilizer use, which was reported to be the critical point in terms of emissions and environmental impacts for the recovered fertilizers, [221] the impact for the Scenario RF (i.e., 3999 $kgCO_{2equiv}$) was only slightly higher than that for the Scenario SF (i.e., 3966 $kgCO_{2equiv}$) because of the higher energy consumption needed for digestate distribution into the soil than that required for urea and other mineral fertilizers distribution (Scenario SF).

From the data reported above, it was derived that the total net impact measured for the production and use of RF was of 3354 $kgCO_{2equiv}$, with this figure being lower (-30%) than that calculated for the Scenario SF, i.e., 4800 $kgCO_{2equiv}$ (Table 4).

GHG impacts were due above all to direct emission of N_2O coming from nitrogen dosed to the soil as fertilizers, with the GHG coming from biogas burning and mass transportation playing only a minor role. The impacts measured for this gas were the same for the two Scenarios studied, since the measured N_2O emissions were not significantly different from each other (Table S4).

Results of this work appear more interesting if it is considered that to add an equal quantity of efficient N to the two Scenarios, much more N was added to the soil in the Scenario RF, i.e., total N of 370 kg ha^{-1} (370 kg ha^{-1} × 0.5 = 185 kg ha^{-1}) (Table S3) than in the Scenario SF, i.e., 185 kg ha^{-1} of N, suggesting that only the efficient (mineral) fraction of total N was responsible for N_2O emission, since these two figures were identical for the two Scenarios studied (i.e., total mineral N dosed of 185 and 185 kg ha^{-1} of N for Scenarios RF and SF, respectively) and that organic N (contained in the digestate) appeared not to additionally contribute at to emissions.

This result was consistent with the high biological stability of the digestate, measured by potential biogas production (BMP) (Table S1), that was even lower (i.e., with higher biological stability) than those reported for well-matured composts, [222]leading to null or a very low rate of mineralization of the organic N in short-medium time. The biological stability of the organic matter has recently been reported to play an important role in defining N mineralization in soil.

Tambone and Adani [223] reported that mineral N produced during organic substrate incubation correlated negatively with CO_2 evolved during soil incubation, i.e., the more stable was the substrate, the less C (and N) mineralization occurred.

In this work, CO_2 and CH_4 measurements carried out directly on plots during the cropping season (Table S4) indicated the absence of differences in C emission for soil fertilized with synthetic fertilizers and digestate but also with the control (no fertilizers added) confirming that organic matter added with digestate was stable, contributing to restore soil organic matter.

The increase of total organic carbon (TOC) in soil treated with digestate after 3 years of fertilization, compared to soil fertilized with mineral fertilizers, seems to confirm this fact (TOC increased after 3 years from 10.3 ± 0.6 g kg^{-1} dry weight (dw) to 12.3 $\pm$ 0.4 g kg^{-1} dw, differently from the mineral fertilized and unfertilized plots that did not show any increase) (unpublished data).

Results obtained in this work differed from those of previous studies that reported higher emissions of N_2O when recovered fertilizers (digestate) replaced mineral fertilizers. [203]

Nonetheless, in that case, N_2O emissions were assumed (not measured directly) to be of 1% of the total N from mineralization, mineral fertilizers, digestate, and existing crop residues; in addition, no data regarding the OM quality of digestate (potential N mineralization), i.e., biological stability, were reported. It can be concluded that N_2O emissions depended on available N (mineral) plus the easily mineralizable fraction of the organic N, which depended, in the first instance, on the biological stability of the organic substrate, so that this parameter becomes important for a rough estimation of the potential N_2O emission.

This result was in contrast with that reported in the literature which indicated a direct proportionality between the total amount of nitrogen supplied and N_2O emissions, [213,224]) without any specification of N type, i.e., organic vs mineral N and organic matter stability responsible for potential N mineralization. We consider that this approach could lead to a misinterpretation of the real impacts of recovered organic fertilizers that need, as already discussed, to be better characterized.

Ammonia emissions represent another important issue in determining environmental impacts when using fertilizers. The full-field approach indicated that there were no differences in ammonia emissions between Scenario RF and Scenario SF (Table S4) thanks to the digestate injection that resulted in a strong mitigation in ammonia emissions in comparison with superficial spreading, [194] as also confirmed by the literature. [193]

The low ammonia emissions did not increase N_2O emission, as already discussed, in contrast with what has been reported in the literature, i.e., that ammonia emissions abatement led to an increase in N_2O emissions, [225] indicating that a well-stabilized organic substrate and the adoption of an efficient distribution technique allowed containment of both NH_3 and N_2O emissions.

The high biological stability of the digestate, providing for low organic matter mineralization, limited, also, the NO_3^- leaching for the Scenario RF, which was, according to the data measured directly at the full field during the crop season, not significantly different from that measured for the Scenario SF (Table S4).

Other Impacts

The identical N_2O emissions reported for the two Scenarios studied led, also, to similar stratospheric ozone depletion impact, since the emissions of ozone-depleting substances (ODSs) are mainly due to direct N_2O emissions from fields. Ionizing radiation quantified the emission of radionuclides in the environment that may be due to nuclear activity, but also to fuel burning.

The Scenario RF achieved a total negative impact because of the production of renewable electricity that compensated for the other emissions caused by transport (transport of sludge to the AD facility), digestate handling, and distribution. Considering just the fertilizer use, the measured impact was higher for the Scenario RF than that for the Scenario SF, i.e., 9.7 vs 4.5 kBq Co-60$_{equiv,}$ (Table 4).

High water content and low-nutrient concentration for digestate, leading to more energy consumption for its distribution than for synthetic mineral fertilizers, were responsible for the higher impact.

The categories ozone formation (human health and terrestrial ecosystem) that quantified the potential molecules leading to the formation of ozone as NO_x equivalent [218] were two of the six categories reported to be higher for the Scenario RF than the Scenario SF, the main contributor to this category being the biogas combustion for electricity production (Figure 10a).

Less important, i.e., about 10%, was the impact due to direct emissions in the field, i.e., distribution of digestate (fuel machinery) and distribution of ammonium sulfate and NO_x direct emissions from land.

Figure 10

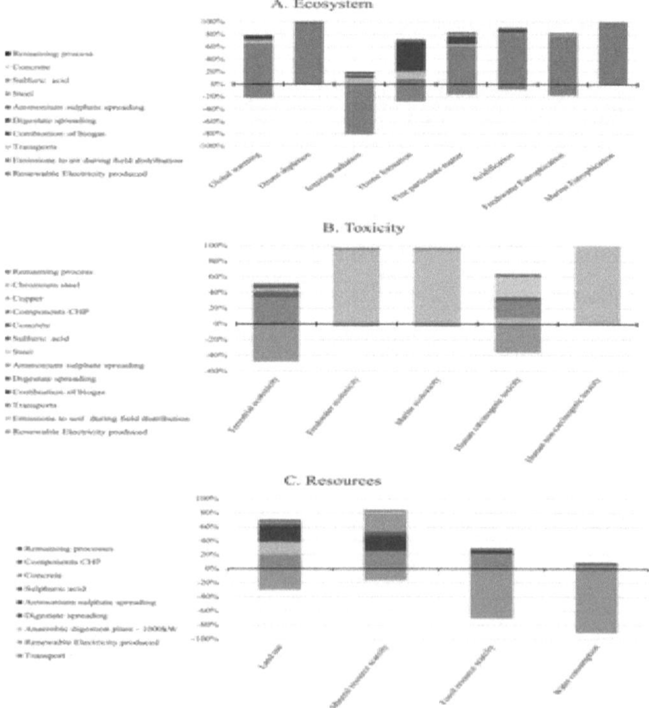

Figure 10. Process contribution to the impact categories of the Scenario RF, focusing on the ecosystem (a), toxicity (b), and resources (c). Impact assessments were calculated according to the ReCiPe 2016 midpoint (H) V 1.03 method and data reported as percent of the total impact.

Impact due to fine particulate matter formation was almost identical for the two Scenarios (Table 4). This result was because this impact was generated mostly by ammonia emissions during field fertilization, which was similar for the two Scenarios investigated (Table S4). Particulate matter due to biogas burning in the CHP unit (producing both heat and electricity), fuel combustion for sludge transport to the plant, and digestate field distribution were balanced by credits due to renewable energy produced, determining only a slightly lower value than that calculated for the Scenario SF.

Terrestrial acidification, which is related to nutrients supplied, i.e., deposition of ammonia, nitrogen oxides, and sulfur dioxide in acidifying forms, displayed similar values for the Scenario RF and Scenario SF (Table 4).

Scenario RF had a slightly higher impact due to fertilizer distribution because of NO_x emissions related to the greater use of machinery necessary for the distribution of digestate. Previous studies reported opposing results, i.e., an increase in potential acidification when N mineral fertilizer was replaced by digestate. [203, 226], On the other hand, when the use of proper timing and distribution techniques were considered, previous LCA results were in line with those of this work. [227,228]

Freshwater and marine eutrophication deal with the increase of nutrients (namely P and N), leading to excessive primary productivity and finally biodiversity losses. Freshwater eutrophication (expressed as P equivalent) displayed a higher value for the Scenario RF than the Scenario SF because the total amount of P brought to the soil by digestate was greater than the crop requirement and so higher than P dosed in the Scenario SF.

Phosphorus overdose depended on the N/P ratio that determined an excess of P when dosing the correct amount of efficient N required by a crop (Table S3). N/P ratio imbalance is well known and documented for animal slurries and digestates, [229] and it is even more accentuated in the case of digestates produced by sewage sludge, in which the previous wastewater purification process mainly determines an accumulation of P, while the denitrification processes displace part of the nitrogen. [230]

For marine eutrophication, the impact measured for the two Scenarios was equivalent, as the N leached assessed in full-field trials was recorded as equal for the two Scenarios studied (Table S4, Supporting Information).

Midpoint Results of Impact Categories Related to Human Health Protection

The inclusion of toxicity categories (USEtox) (Table 4) in the ReCiPe 2016 methodology allowed us to better focus the impacts of the production and use of fertilizers when compared with previous work done that considered only the main agricultural-related indicators, such as global warming potential, eutrophication, and acidification. [203,228]

The use of fertilizers determined a higher impact for the Scenario RF than the Scenario SF for the toxicity categories, i.e., Freshwater and marine ecotoxicity and human noncarcinogenic toxicity, because of heavy metals (HM) (above all Zn) supplied to soil with digestate. This figure has already been highlighted in literature for other organic fertilizers (pig slurries) because of their very high Zn and Cu contents. [231,232]

In particular, the amount of Zn applied to the soil with the digestate corresponded to 3.8% of that present in the 15 cm of surface soil, but after 3 years of experimentation, no differences were observed in soil Zn content (Table S5). Nevertheless, analyzing grains, higher Zn content was revealed for plot amended with digestate (Table S5) although the same grain production was measured (Table S6). However, this content was in line with those reported in the literature for both maize grain and other cereals (i.e., rice and wheat). [233]

Further effort should be made to decrease impacts, reducing HM in sewage sludge by selecting the cleanest ones. The terrestrial ecotoxicity impact was mainly generated during the fertilizer production (Table 4); in particular, for the Scenario RF, the impact was due above all to the transport of sludge to the AD plant (Figure 10b), while for the Scenario SF, it was the N fixation process (ammonia steam reforming) that determined the impact.

Nevertheless, the Scenario RF benefitted from the production of electricity, significantly reducing the impacts. Finally, the category human carcinogenic toxicity also showed a better environmental outcome for the Scenario RF than the Scenario SF, thanks to the credits from the production of renewable energy (Figure 10b).

Midpoint Results of Impact Categories Related to Resource Scarcity Protection

The use of both renewable energy (biogas) and recovered material (sewage sludge) to produce fertilizers (digestate and ammonia sulfate) led, also, to high efficiency in terms of land use, mineral resource use, fossil resources, reducing, until negative, these impacts (Table 4).

Single End Point Indicator

The single end point indicator provided by the ReCiPe method allows one to view the normalized and weighted impacts in a synthetic manner and is divided into the three areas of protection, i.e., ecosystem, toxicity, and resources (Figure 9). The Scenario RF was significantly better than the Scenario SF, and in particular, the indicators showed for the Scenario RF, not only an impact reduction but also the prevention of impact in the areas of protection of resources and human health, as previously reported. [198,234–237]

Further Scenarios Reducing Environmental Impacts in Producing and Using Renewable Fertilizers

Life cycle assessment is a powerful tool for describing impacts due to fertilizer production and use, highlighting positive and negative effects for renewable fertilizers vs synthetic mineral fertilizers in a real case study. However, LCA is also a potent tool to design potential Scenarios in terms of environmental impacts, from which to learn how to improve productive processes and further reduce environmental impacts. This process can be done by observing in detail impact categories and the contribution of each process activity to the category impact to find solutions by combining individual technologies. [238]

The results discussed above indicate that the recovery of sewage sludge producing renewable fertilizers by AD allowed environmental benefits when the renewable fertilizers produced were used correctly and by efficient timing in substituting for synthetic mineral fertilizers, suggesting that the application of the Circular Economy in agriculture in terms of fertilization resulted in a win–win approach, which makes it more sustainable. However, as for all productive processes, impacts remain, and they cannot be nullified completely but only further reduced.

The detailed observation of every single impact, divided for impact categories and activities affecting each impact (Figure 10), allowed us to understand what are the more important factors in determining impacts. Emissions to air during field distribution of fertilizers (i.e., NH_3 and N_2O emission) seemed to affect greatly the ecosystem and human toxicity categories as they interacted with many impact subcategories (Figure 10a,b).

Therefore, reducing air emissions allows the further reduction of an ecosystem and human impacts because of renewable fertilizer production and use. Digestate and ammonium sulfate produced by the plant studied in this work were used correctly following the best practice, i.e., digestate and ammonia injection, while the digestate was characterized by high biological stability, avoiding N mineralization and nitrate leaching.

The strong impact reduction obtained by substituting synthetic mineral fertilizers with renewable fertilizers (Table 4 and Figure 9) confirmed this virtuous approach. Nevertheless, already stated, LCA can help in optimizing processes, further reducing impact.

Nitrogen dioxide emissions have been reported to be greatly reduced using nitrification inhibitors (NI). [239, 240] From the literature, it was calculated, on average, that the use of NI allowed a reduction of 44% in total N_2O emissions, [241] further reducing total Scenario RF impacts (Scenario RF_1), with reference to ecosystem and human health impacts (Figure 11), if these data are implemented in the LCA. The modeling of this Scenario considered just the addition of NI to the soil reducing N_2O emission (data from literature). [237]

The production (dicyandiamide) and distribution of the nitrohinibitor were considered negligible because of the very limited amount of product used (ca. 7 kg ha^{-1}). In doing so, all of the data describing the Scenario remained the same as the original one (Scenario RF).

Figure 11

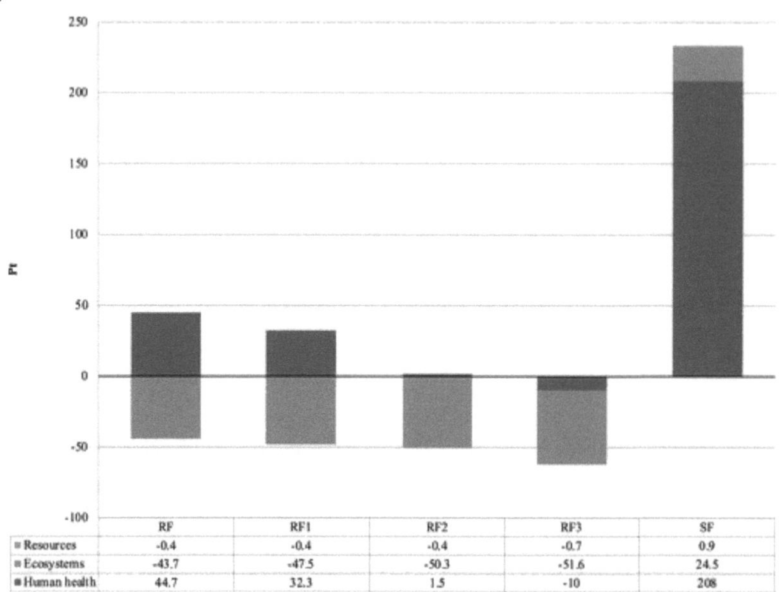

	RF	RF1	RF2	RF3	SF
▪ Resources	-0.4	-0.4	-0.4	-0.7	0.9
▪ Ecosystems	-43.7	-47.5	-50.3	-51.6	24.5
▪ Human health	44.7	32.3	1.5	-10	208

Figure 11. Comparative environmental results (Ecopoint—Pt) for the Scenario RF (recovered fertilizers), Scenario RF$_1$ (RF + nitro inhibitor), Scenario RF$_2$ (RF + nitro inhibitor + anchor), Scenario RF$_3$ (RF + nitro inhibitor + anchor + biomethane for transportation), and Scenario SF (synthetic fertilizers). Impact assessment was calculated according to the ReCiPe 2016 end point (H) V 1.03 method.

On the other hand, total ammonia emitted during digestate distribution can be reduced by optimizing the injection system. Preliminary data coming from work performed at full scale at the AD plant studied in this work indicated that by modifying the distribution equipment, i.e., Vervaet Terragator equipped with flexible anchors and a roller postposed to the anchors, allowed a reduction of ammonia emission of 44% (data not shown).

The future integration of this practice will allow a further reduction of impacts, as shown in Figure 11 (Scenario RF_2). Because the anchor system was applied to the digestate distribution system already in use, the only change in the Scenario modeling was referred to the emission of ammonia measured.

Another important activity that plays an important role in determining impact is transport. Transport affected a lot the terrestrial ecotoxicity (Figure 10b) and, although much less severely, many other subcategories within ecosystem and resources categories (Figure 10a,c) because of the fossil fuel used. Today, in the EU, anaerobic digestion represents a well-consolidated bioprocess treating organic wastes and dedicated energy crops, producing biogas/biomethane. [242]

In the Lombardy region alone, about 580 AD plants are operating producing biogas and now are starting to produce biomethane. [243, 244] Recently, a particular interest has been devoted to liquid biomethane (Bio-LNG) as a substitute for fossil fuels in truck transportation, [245] and the first plants have started operating in Lombardy region, very close to the AD plant studied in this work.

A new Scenario was modeled (RF3) assuming the biogas production from organic wastes (OFMSW and sludge), the purification and compression of biomethane, and the transport by 30 ton trucks and average consumption of fuel equal to 0.34 kg LNG per kilometer traveled. [246]Emissions from trucks were recalculated accordingly.

Assuming an ability to substitute all fossil fuels with Bio-LNG produced from the organic fraction of municipal solid waste (Table 3) for transportation, a further strong impact reduction was obtained, nullifying completely the environmental impacts due to production and use of recovered fertilizers (Scenario RF_3) (Figure 11).

Fertilisers provide nutrients for plants. Nutrients needed in the largest quantities in agriculture are nitrogen, phosphorus and potassium.

The adverse publicity given by the media to agriculture's role in polluting the environment may make farmers feel guilty about using fertiliser. However, reducing fertiliser input can lead to reduced plant growth which can aggravate problems such as soil erosion.

It is important for you to be aware of the effects of fertilisers and to use them carefully, but it is also important that everyone realises that agricultural fertilisers are not the main sources of environmental pollution. If you apply fertilisers sensibly, so that plants use all the nutrients and none are leached, there is little opportunity for pollution.

Nutrients

Nitrogen

On farmed land, most nitrogen is in organic matter which must first be mineralised by soil microbes into ammonium or nitrate to be used by plants. Nitrate is easily leached from soil and so presents the most opportunity for pollution.

Phosphorus

Phosphorus is a very stable element and moves only 1–5 mm from where it is spread. It binds quickly with soil minerals, so is unlikely to leach through soil except under high rainfall in very sandy soils. It is mainly lost from the soil by erosion when soil particles holding the phosphorus are blown or washed away. For this reason fertiliser phosphorus is unlikely to be a major contributor to phosphate pollution of waterways, unless erosion occurs.

Potassium

Potassium is taken up by plant roots very rapidly and is not used in great quantities, so represents little environmental threat. Bananas need large quantities of potassium and care needs to be taken to apply it in small amounts, often, so that the plant can use all of it.

Environmental hazards

Groundwater pollution

Nitrate leaching through the soil can present a serious health hazard and contributes to soil acidification. When high rates of nitrogen are used or where clover grass pastures fix substantial nitrogen, especially on sandy or permeable soils, inevitably some nitrate is leached and may enter groundwater if there is a watertable. If this groundwater is used for domestic supplies, the leaching presents a serious health hazard.

Eutrophication

Eutrophication is the enrichment of water by the addition of nutrients. The extra nutrients encourage the growth of algal blooms, particularly in stagnant water. Blue–green algae may produce toxins poisonous to animals, including humans. For this algae to grow, phosphorus must be present in the water above a certain level. Phosphorus may be introduced into waterways in run-off from pasture, forests and fertilised land, and in drainage from irrigated land and urban areas.

These sources, representing most of the total run-off, normally contribute low concentrations of phosphorus and are referred to as *diffuse* or *non-point sources*. *Point* sources, such as sewage effluent and drainage from dairies and feedlots, contribute smaller flows but contain much higher concentrations of phosphorus. These are frequently found to be the sources for most of the phosphorus found in waterways.

Soil acidity

There are three major acidifying processes in NSW agricultural systems:

1.   addition of nitrogen to the soil by fertiliser or fixation of atmospheric nitrogen, followed by loss of nitrate from the soil due to leaching or run-off

2.   production of organic acids from decomposing organic matter

3.   removal of alkaline products such as hay from the soil.

Contrary to popular belief, superphosphate does not cause soil acidification.

You can take several actions to lower acidification rates in your soil.

•   Use less acidifying nitrogen fertilisers: for example, use urea rather than ammonium sulfate.

•   Incorporate stubbles into fallow to minimise net nitrification.

•   Sow early to maximise the opportunity of the crop to recover soil nitrate.

•   Use perennial deep-rooted plants able to rapidly absorb mineralised nitrate at the start of the growing season and maintain low soil nitrate levels throughout the year.

•   Use deep-rooted crops.

•   Minimise water percolation below the root zone.

•   Avoid excessive irrigation.

•   Minimise removal of product from the soil. To prevent acidification you need to apply 55–60 kg of lime for every tonne of lucerne or clover hay removed; 35 kg of lime per tonne of grass hay removed; 22 kg of lime per tonne of cereal hay removed; and 3 kg of lime per tonne of cereal grain removed.

•   Minimise manure removal from pastures, preferably leaving manure where the animals graze.

•   Feed hay on the paddocks where it is cut.

•   Use cropping rotations to minimise excessive accumulations of soil organic matter under pasture.

Cadmium

Cadmium is present in tiny amounts (less than 0.5 mg/L) in the soil, and in larger amounts in rock phosphate. Plant uptake of cadmium is small, but when plants containing cadmium are grazed by livestock, the cadmium accumulates in offal and may reach very high concentrations. This is a severe problem on the sandy grazing soils of South Australia and Western Australia, but is not a problem on soils with even a low clay content.

Fertiliser guidelines

Superphosphate

- Don't topdress dams, streams or swampy areas.

- Don't topdress bare ground.

- Maintain good groundcover around dams and streams.

- Avoid topdressing when heavy cyclonic rain is expected.

Nitrogen

- Use the least acidifying fertiliser you can afford.

- Apply it in small amounts frequently rather than all at once, to minimise nitrate leaching.

Effect of addition mineral, organic and bio-fertilizers on nitrogen, phosphorous, potassium concentration and protein of corn crop (Zea mays L.)

Soil fertility and plant nutrition is an ongoing program and operations of modern management to increasing productivity or improving the quality of crops, and the addition fertilizers either to increase soil fertility or to compensate for the lack of available nutrients or maintain a good balance between nutrients, especially nitrogen, phosphorous and potassium [247].

Most of Iraqi soils, suffer from a lack of nitrogen, phosphorus and potassium, this means that fertilization increases growth and living mass, and this is the reason behind the importance of mineral fertilization in agricultural production, and although nitrogen is abundant in Earth, but mostly of it is not available for living organisms and the amount of nitrogen in Soil surface and readiness for living organisms does not exceed 0.03% [248].

In alkaline and calcareous soils there are phosphorous and potassium in compounds, as well as phosphorus associated with clay particles and organic matter, and subjected to many reactions that reduce its readiness in soils such as adsorption and sedimentation on the surface of minerals, soil particles, organic matter, carbonate minerals and the interaction of phosphorus with Calcium ions, which requires the enhancers adding, to increase their solubility in the soil and itsavailability to plant [247].

The importance of potassium comes from its important role in many biological physiological processes and in stimulating many enzymatic reactions in plants, especially the transport and storage of materials represented and water relations within the plant [249, 250].

The use of organic fertilizers is important to improving soil chemical properties, fertility and contributes with mineral fertilizers to increasing the productivity of crops [251, 252]. Humus is the final and most stable product of organic matter decomposition in soil and its decomposition is very slow [253].

Biological fertilization is used instead of chemical fertilizers to provide nutrients for plant with a compensation rate of more than 25%, as well as its role in reducing environmental pollution problems and working on agriculture sustainability [254]. Integrated fertilization was used including biological fertilizers along with mineral and organic fertilizers to increase nutrients availability for plant and its impact on productivity [255].

[256] Indicated that many crops responded to biological fertilization by using PGPB (Plant Growth Promotion Bacteria) in recent years, which encouraged the manufacture of bio fertilizers. Corn is an important grain crops due to its versatility as it is used in food and in many industries such as oil and biofuels, as well as using of dry plant material for animal fodder. Therefore, the use of mineral fertilizers, including nitrogen, phosphorus and potassium, is required to increase the growth and productivity of the plant [257, 258].

The nutrient is chemically available of plant, if it is available in a form that is capable of being absorbed by the plant, and since Iraqi soils are calcareous soils, so research aims to provide available elements from various fertilizer sources and to maintain soil fertility and productivity.

the levels of mineral fertilizer gave significant differences of nitrogen concentration and protein, as the level of P2 exceeded an increase 32.16, 33.83 and 33.73% compared to the level of P0 in dry matter, grains and protein, respectively. the increase in the levels of organic fertilizer led to a significant increase of nitrogen concentration and protein, where the level M1 exceeded with an increase 55.48, 58.98 and 58.38% compared to the level M0 in dry matter, grains and protein, respectively.

As for the biological fertilizer levels achieved a significant differences, and the best was the level B1, which gave the highest nitrogen concentration and protein percentage, with an increase 38.55, 42.10 and 42.91% compared to the level B0 in each of dry matter, grains and protein respectively.

The interference between the organic and the biological fertilizer was the only significant effect on nitrogen concentration in dry matter, grains and protein, as the best level was M1B1, which gave the highest value, with an increase 111.94, 120.91 and 121.46% compared to M0B0 in dry matter, grains and protein, respectively. The results show that there are significant differences for the triple interference, as the level P2M1B1 exceeded an increase 141.86, 145.33 and 144.68% compared to the level P0M0B0, which gave the lowest value for nitrogen concentration in dry matter, grains and protein respectively.

The addition of mineral fertilization has a role in increasing the absorption of nitrogen in the plant, especially NH4, in order not to compete with the phosphate ions H2PO4 and HPO4 for nitrogen on the sites of absorption and thus increase the concentration and absorption of nitrogen in the grains, which increased the percentage of protein in the grains, as well as its essential role in the metabolic reactions of all living organisms as all Processes in the cell use adenosine diphosphate (ADP) or triphosphate (ATP) as an energy source such as respiration and reproduction, photosynthesis, and protein biosynthesis [247].

The organic matter and its decomposition in the soil leads to an improvement in the physical and chemical properties of the soil and an increase in the elements available for absorption, especially nitrogen, as well as its nitrogen content, which leads to an increase in its availabilityin the soil and thus an increase in its absorption by the plant, which is reflected positively on the concentration of nitrogen in the grains, thus increasing the percentage of protein In grains.

that the levels of mineral fertilizer gave significant differences of phosphorus concentration, as the level P2 exceeded an increase 36.84 and 25.92% compared to the level P0 in dry matter and grains, respectively. The results showed the increasing of organic fertilizer levels led to a significant increase of phosphorus concentration, where the M1 gave the highest concentration with an increase 64.70 and 44.00% compared to M0, which gave the lowest phosphorus concentration in dry matter and grains, respectively.

The addition of biological fertilizer had a significant effect, and the level B1 was superior in increasing of phosphorus concentration, with an increase 50.00 and 30.76% compared to the level B0 in both dry matter and grains, respectively. The results showed that there are not significant differences in double interference, except for interference between organic and biological fertilizers in phosphorus concentration of dry matter and grains, Significant differences were observed, and M1B1 was the best, which gave the highest value of phosphorous concentration, with an increase 161.53 and 86.36%, compared to the level M0B0 in dry matter and grains, respectively.

The results showed that there are significant differences of triple interference, as P2M1B1 level gave the highest value for phosphorus concentration with an increase 208.33 and 109.52% compared to P0M0B0 in dry matter and grains respectively.

The increased levels of mineral fertilizer added led to an increase in the availability of phosphorus in the soil that could be absorbed by the roots, which in turn was reflected in the concentration of phosphorus in the plant, as phosphorus strengthens and activates the root system of plants and thus increases the absorption of nutrients, including phosphorous [250], In addition to added organic fertilizer containing phosphorus and its role of increasing the availability of phosphorus when it is dissolved in the soil [256].

In addition to the presence of phosphate-dissolving bacteria in the root environment, it contributed to increasing the phosphorous availability in the soil from its various sources, as it contributed to dissolving various calcium phosphates such as mono, di- and triple calcium phosphate insoluble to dissolved by the effect of phosphorus-dissolving organisms and thus encouraged the absorption of phosphorus by the plant [252].

the levels of mineral fertilizer gave significant differences of potassium concentration, as the level P2 exceeded an increase 26.53 and 23.32% compared to P0 level in dry matter and grains, respectively. The results showed that increasing the levels of organic fertilizer led to a significant increase of potassium concentration, as M1 level outperformed it ingiving it the highest rate with an increase 42.97 and 37.88% compared to M0 level in dry matter and grains, respectively.

As for the levels of biological fertilizer achieved significant differences, and the best was B1, which gave the highest rate, with an increase 11.78 and 25.95% compared to B0, which gave the lowest rate in dry matter and grains respectively. The results showed that the interference between organic and biological fertilizers in potassium concentration in dry matter gave significant differences, and the best was M1B1, which gave the highest value, with an increase 85.44% compared to M0B0 in dry matter.

The results also showed that there are significant differences for triple interference, as the level P2M1B1 exceeded of potassium concentration with an increase 112 and 92.96% compared to level P0M0B0, which gave the lowest value of potassium concentration in dry matter and grains respectively. mineral, organic and biological fertilization has an important role in increasing the availability of nutrients in the soil and increasing the possibility of their absorption by the roots, which led to an increase in the content of these elements, including potassium, As well as the promotional relationship between nitrogen and potassium in absorption by plants.

In addition to the role of organic fertilizer in improving many physical and chemical properties and improving the availability of nutrients, as well as the potassium content of added fertilizer and its role in increasing the availability of potassium in the soil. The role of biological fertilization is also important in improving the level of potassium in the plant and improving its growth through the secretion of growth regulators, the most important of which are oxins, gibberellins, and cytokines that encourage plant absorption of nutrients [253, 254].

Biofertilizer: The Future of Food Security and Food Safety

Global demand for agricultural products is increasing due to the increasing human population [266]. There are already about 7.9 billion people on the planet, and this number is expected to rise, with a projected growth of almost 10 billion in the next 50 years [267-269]. As the world's population continues to increase, so does the demand for food; hence, feeding the current vast population, which will certainly grow with time, is a significant task [270].

To meet the challenges of food scarcity caused by the rise in population, various agricultural alternatives such as the use of chemical or synthetic fertilizers, pesticides, and insecticides have been used to produce crops with high yield within the shortest time possible and to protect them from insects and pest attack during and after harvest [271].

However, the use of these fertilizers and insecticides has raised much public concern about the sustainability, safety, and security of the food supply [7,8]. Studies have shown that there is a significant amount of pesticide residue present in foodstuffs long after they are taken away from farms for human consumption [274]; hence, the need for alternatives such as biofertilizer in ensuring food safety and security [271].

Moreover, synthetic fertilizers that consist of various nutrients such as nitrogen (N_2), phosphorus (P), potassium (K), and sulfur may become harmful if used beyond the required amount [267].

The harmful effects of these fertilizers include the weakening of plant roots, the high rate of disease incidence, soil acidification [275], and eutrophication of ground water and other water bodies [276]. Nutrients such as nitrates leach to groundwater and cause "blue baby syndrome", also called "acquired methemoglobinemia" [267].

The impact of these chemicals will not only affect the present but also future generations. Therefore, there is need to search for eco-friendly approaches such as biofertilizers, which play a major role in sustainable agriculture [267].

Biofertilizers are microorganisms that support the growth of plants by enhancing the nutrient supply to the host plant when given to seeds, plants, or the soil [267-278]. They colonize the rhizosphere or the inside of the plants.

This entails the use of plant growth-promoting microorganisms that participate in a variety of biotic activities in the soil ecosystem in order to make it dynamic and sustainable for crop development. Biofertilizers are widely used to accelerate microbial activities that increase the availability of nutrients that plants can easily absorb.

They increase soil fertility by fixing atmospheric N_2 and solubilizing insoluble phosphates in the soil, resulting in plant growth-promoting chemicals [279]. These biofertilizers make use of the naturally available biological system of nutrient mobilization, which greatly enhances soil fertility and, as a result, crop productivity. [279].

It has been reported that the biofertilizer market is estimated to grow at a compound annual growth rate (CAGR) of 14.0% from 2015 to 2020 and is expected to reach USD 1.88 billion by 2025 [280]. Because of strict regulations on the use of chemical fertilizers, biofertilizers are the most widely used in Europe and Latin America [280].

The words "food security" and "food insecurity" are commonly used in discussions of global conditions and prospects. Food security is defined as the availability and accessibility of safe and nutritious food that fits the dietary requirements of a healthy and active lifestyle.

Food insecurity occurs when people do not have enough access to safe and nutritious food, and thus do not consume enough to live an active and healthy life. This could be due to a lack of food, a lack of purchasing power, or inefficient use of resources at the household level [281].

Another factor that may be responsible for food insecurity may be the depletion of soil nutrients resulting from continuous tillage and the use of chemical or synthetic fertilizers for continuous agricultural production. This have made the soil lose its fertility, and most of the agricultural produce consumed is not safe because of the chemical residues that are left in them.

This review highlights the role of biofertilizers in crop improvement and the production of safe and secure food, the mechanisms of microorganisms in enhancing plant growth, and the various types of organisms used as plant growth-promoting microorganisms.

Biofertilizers

Microbial inoculants, also known as biofertilizers, are organic products that contain specific microorganisms obtained from plant roots and root zones. They have been found to boost plants' growth and yield by 10–40% [281]. These bioinoculants colonize the environment when applied to the rhizosphere and the interior of the plant to promote plant growth [282].

They not only add nutrients to the soil to improve soil fertility and crop yield, but they also protect the plant against pests and diseases. They have been shown to enhance seedling survival, extend the root system's life, eliminate harmful chemicals, and shorten flowering time [276]. Another advantage is that biofertilizers are no longer necessary after 3–4 years of continuous use, since the parental inocula are sufficient for growth and multiplication [283].

Plants require 17 essential elements for effective growth and development. N_2, P, and K are all required in significant amounts [283]. Several microorganisms, including nitrogen-fixing soil bacteria and cyanobacteria, phosphate-solubilizing bacteria, molds, and mushrooms, are routinely utilized as biofertilizers [284]. Similarly, microorganisms that produce phytohormones are used in the production of biofertilizers. They feed the plant with growth-promoting compounds such as indole acetic acid (IAA), amino acids, and vitamins, as well as improving the soil's productivity and fertility while conserving crop yield [285].

Types of Biofertilizers

Biofertilizers are divided into groups based on their functions and mechanisms of action. The most commonly used biofertilizers are nitrogen-fixers (N-fixers), potassium solubilizers (K solubilizers), phosphorus solubilizers (P solubilizer), and plant growth-promoting rhizobacteria (PGPR) [282].

One gram of rich soil can contain up to 10^{10} cfu bacteria, with a live weight of 2000 kg/ha [286]. Cocci (spheres with a diameter of 0.5 m), bacilli (rods with a diameter of 0.5–0.3 m), and spirals with a diameter of 1–100 m are all types of soil bacteria. The frequency of bacteria in the soil is influenced by the physical and chemical properties of the soil, organic matter, and phosphorus concentration, as well as cultural activities.

Nutrient fixation and improvements in plant growth by bacteria, on the other hand, are critical components for accomplishing future sustainable agricultural goals. Microbes also help the ecosystem's numerous nutrient cycles. Table

5 summarizes the classification of biofertilizers based on the type of microbe utilized and the mechanism of action, as well as appropriate examples.

Table 5

Classification of biofertilizers and their mechanism of action.

Biofertilizers	Mechanism	Groups	Examples	References
Nitrogen-fixing	Increase the amount of N_2 in the soil by fixing atmospheric nitrogen and making it available to plants.	Free-living, symbiotic, and associative symbiotic	*Aulosira bejerinkia, Nostoc, Klebsiella, Stigonema, Desulfovibrio, Azotobacter, Anabaena, Clostridium, Rhodospirillum,* and *Rhodopseudomonas Rhizobium, Frankia, Anabaena azollae,* and *Trichodesmium Azospirillum* spp., *Herbaspirillum* spp., *Alcaligenes, Enterobacter, Azoarcus* spp., and *Acetobacter diazotrophicus*	[282]
Phosphorus-mobilizing	Phosphorus is transferred from the soil to the root cortex.	Mycorrhiza	*Arbuscular mycorrhiza, Acaulospora* spp., *Scutellospora* spp., *Glomus* spp., *Gigaspora* spp., and *Sclerocystis* spp.	[282]

Biofertilizers	Mechanism	Groups	Examples	References
	These are bio-fertilizers with a wide range of applications .			
Potassium solubilizing	Produce organic acids that degrade silicates and aid in the removal of metals to solubilize potassium (silicates) ions and make it available to plants.	Bacteria	*B. edaphicus*, *Arthrobacter* spp., *Bacillus*, *Mucilaginosus*, and *B. circulanscan*	[287]
		Fungi	*Aspergillus niger*	
Potassium	They	Bacteria	*Bacillus* spp.	[288]

Biofertilizers	Mechanism	Groups	Examples	References
mobilizing	transfer potassium from the soil's inaccessible forms.			
		Fungi	*Aspergillus niger*	
Phosphorus solubilizing	To dissolve bound phosphates, they secrete organic acids and lower soil pH by converting insoluble forms of P in the soil into soluble forms.	Bacteria, fungi	*Pseudomonas striata, Bacillus circulans, Bacillus subtilis, Penicilium* spp., *B. polymyxa, Agrobacterium, Microccocus, Flavobacteri um, Aereobacterium. Aspergillus awamori, Penicillum* spp., and *Trichoderma* spp.	[282]
	Sulfur is oxidized to sulfate,	Sulfur-oxidizing	*Thiobacillus* spp.	[289]

Biofertilizers	Mechanism	Groups	Examples	References
	which is the usable form for plants.			
Micronutrient	Protons, chelated ligands, acidification, and oxidoreductive systems can all be used to dissolve zinc.	Zinc-solubilizing	*Pseudomonas* spp., *Mycorhiza*, and *Bacillus* spp.	[290]
Plant growth-promoting	Produce hormones that encourage root growth, increase nutrient availability, and boost crop yields.	Plant growth-promoting rhizobacteria	*Agrobacterium, Pseudomonas fluorescens, Arthrobacter, Erwinia, Bacillus, Rhizobium, Pseudomonas* spp. *Enterobacter, Streptomyces,* and *Xanthomonas*	[282]

Plants are exposed to diverse microorganisms in their natural habitat, including bacteria, fungi, algae, and protozoa. The majority of these microorganisms occur in the soil's rhizosphere in various types of association, some as free-living organisms, while others associate with plant roots or even live within root or shoot tissues as endophytes [291,292].

In the instance of a symbiotic relationship with nitrogen-fixing bacteria in the root nodules of leguminous plants, these connections may be advantageous to the plant, while others may be parasitic, pathogenic, or have no known effect on plant growth or development [292].

Microorganisms that promote plant growth are involved in a variety of biotic activities in the soil ecosystem to keep it dynamic and sustainable for crop production [293].

They colonize plant roots competitively and improve plant growth through a variety of mechanisms, including phosphate solubilization [294]; nitrogen fixation [295]; production of indole-3-acetic acid (IAA), siderophores [296], 1-amino-cyclopropane-1-carboxylate (ACC) deaminase, and hydrogen cyanate [297]; degradation of environmental pollutants; and the production of hormones, antibiotics, and lytic enzymes [298].

Furthermore, some plant growth-promoting rhizobacteria may be able to stimulate additional particular plant growth-promoting properties, such as heavy metal detoxification, salinity tolerance, and biological control of phytopathogens and insects [299].

Desulfovibrio, Rhodospirillum, and *Rhodopseudomonas* are examples of beneficial microbes that create symbiotic partnerships with plants, exchanging carbon-based photo-assimilates for minerals ingested by the microbe. Plant and soil biologists have paid extensive attention to beneficial symbiotic microorganisms in recent years, with major goals being the identification and adoption of new, environmentally beneficial lines of plant growth-promoting (PGP) microorganisms. Plant growth stimulators have also been found in other biostimulators, such as those found in seaweed extracts or decomposed vegetation [291].

Mechanisms of Action of Plant Growth-Promoting Rhizobia

There are different mechanisms by which plant growth-promoting rhizobacteria stimulate the growth of plants. They are widely classified as direct or indirect mechanisms [300]. Moreover, depending on their association with the plants, plant growth-promoting rhizobacteria are grouped as both symbiotic bacteria and free-living rhizobacteria [300].

Examples of plant growth-promoting bacteria include the free-living bacteria which form distinct symbiotic relationships with plants, endophytic bacteria which colonize some portions of plant tissue, and cyanobacteria [300]. Despite the differences that exist among the bacteria, they all show a similar type of mechanism while promoting bacterial growth [300].

The bacteria may use one of two methods to promote plant growth by (i) directly by improving resource acquisition or changing the plant's hormone levels, or (ii) indirectly by lowering the inhibitory effects of various pathogenic agents on plant growth and development (Figure 12).

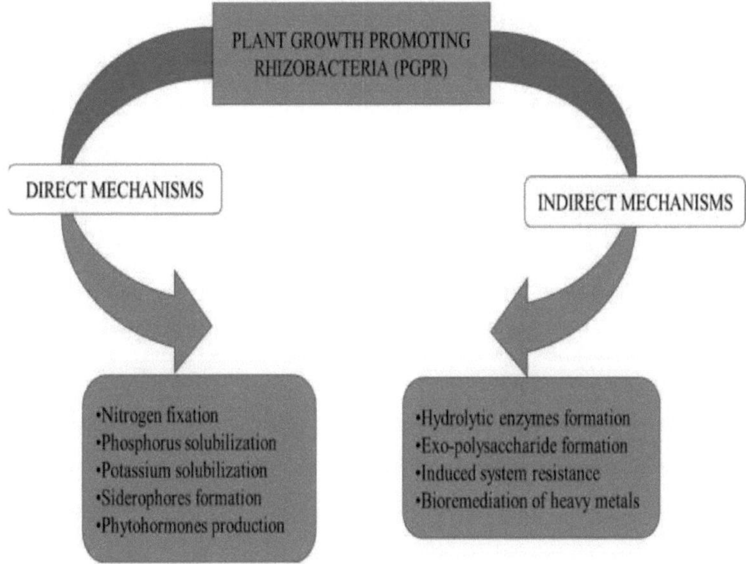

Figure 12

Mechanisms of plant growth-promoting rhizobia.

Direct Mechanisms

Facilitating Resource Acquisition

Biofertilizer aids in nitrogen fixation, iron sequestration, and phosphate solubilization, allowing plants to use these complex organic molecules.

One of the most important nutrients for plant growth is nitrogen. Although our atmosphere contains around 80% gaseous nitrogen, green plants are unable to utilize it directly [301]. Biological nitrogen fixation is the conversion of atmospheric nitrogen to ammonia by soil-borne microbes. About 175×10^6 tons of nitrogen are fixed globally each year by nitrogen-fixing bacteria [302].

Biological nitrogen fixation is a critical component of microbial activities. Only prokaryotes, which can be symbiotic or free-living in nature, are able to produce the nitrogenase enzyme to fix nitrogen biologically. The enzyme nitrogenase catalyzes biological nitrogen fixation. Some soil bacteria and blue-green algae can convert nitrogen from the air into ammonia in their cells. Diazotrophy, or nitrogen fixation, is the process of nitrogen reduction [294,302-305]. N-fixers, also known as diazotrophs, are microbes that reduce atmospheric nitrogen. Plants can directly utilize the ammonia produced during nitrogen fixation.

The *Rhizobiaceae* (α-proteobacteria) are a family of symbiotic N_2-fixing rhizobacteria that live in a symbiotic association with leguminous plant roots. This relationship necessitates a complicated interaction between the host and the symbiont, which leads to the creation of nodules that house the rhizobia as an intracellular symbiont [306].

The   rhizobia include *Rhizobium, Bradyrhizobium, Sinorhizobium, Azorhizobium,* and *Mesorhizobium* as a group. Rhizobacteria that fix nitrogen in non-leguminous plants are known as non-symbiotic rhizobacteria. They are also known as diazotrophs, and they can create a non-obligate relationship with their hosts [307].

The nitrogen fixation process is carried out by nitrogenase, a complex enzyme structure that includes dinitrogenase reductase, which has iron (Fe) as a cofactor, and dinitrogenase, which has iron (Fe) and molybdenum (Mo) as cofactors [304]. In Figure 13, dinitrogenase reductase produces electrons and uses them to decrease N_2 to NH_3 [308].

Mo-nitrogenase, V-nitrogenase, and Fe-nitrogenase are three different nitrogenase complexes based on changes in the cofactor of dinitrogenase [305,309,310]. N_2 fixation genes, also known as *Nif* genes, are found in both symbiotic and free-living nitrogen-fixing microorganisms [309]. *Nif* genes are structural genes involved in Fe–protein activation, Fe–Mo cofactor biosynthesis, electron donation, and serve as regulatory genes required for enzymatic synthesis and activity [310].

Despite being a negative regulator of *Nif* gene expression, oxygen is required for *Rhizobium* sp. bacteroid respiration [311]. Because bacterial leghemoglobin has a high affinity for oxygen, it can keep the enzyme active even in the absence of oxygen (Figure 13). To efficiently pursue the nitrogen fixation process, sufficient O_2 supply to the bacteroid for respiration must occur concurrently with prevention of the O_2 supply to the nitrogenase enzyme complex.

The simplest way to accomplish this objective is to use genetic engineering to introduce bacterial hemoglobin (Hb) that binds O_2 to the rhizobacteria [312]. Following this strategy, it was discovered that after transforming *Rhizobium etli* with the Hb gene of *Vitreoscilla* sp. (a Gram-negative bacterium), the rhizobial cells had a two- to threefold faster respiration rate than non-transformed rhizobial cells [295].

Because *Vitreoscilla* sp. has Hb-producing genes, inserting this gene into rhizobial cells resulted in Hb production in the transformed cells. Despite the low availability of O_2, the Hb generated in this way could bind to it with a high affinity. When the altered *Rhizobium* was inoculated to bean plants, the plants had 68% greater nitrogenase activity than plants inoculated with wild-type *R. etli*. The resulting seeds had a 25–30% increase in leaf content and a 16% rise in nitrogen content as a result of this change [313].

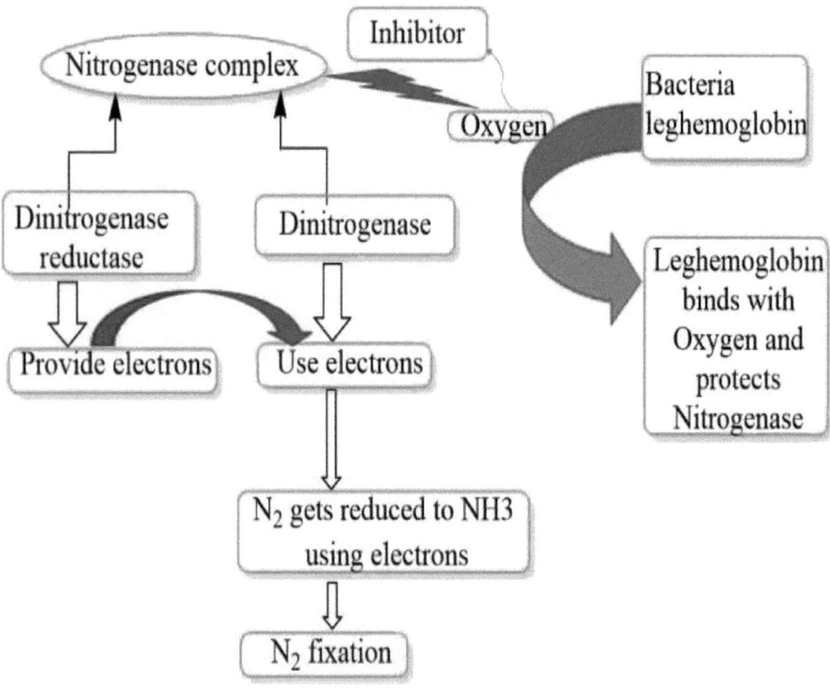

Figure 13

Plant growth-promoting rhizobacteria's molecular N_2 fixing mechanism. The nitrogen fixation process is carried out by the nitrogenase enzyme complex, which comprises dinitrogenase reductase and dinitrogenase. Dinitrogenase reductase produces electrons, which dinitrogenase uses to convert N_2 to NH_3. Because the enzyme complex can attach to O_2 and become inactive, oxygen is a powerful inhibitor. Bacterial leghemoglobin, on the other hand, has a higher affinity for oxygen and hence binds to free oxygen more effectively. As a result, the presence of leghemoglobin protects the nitrogenase enzyme complex from oxygen, keeping it active [281].

Nodule development is another key component of *Rhizobium*. To accommodate the symbiotic bacteria *Rhizobium*, most bean plants generate root lateral organs de novo, known as "root nodules". Symbiotic bacterial infection of the legume plant stimulates the creation of new organs, such as nodules, by altering the fate of differentiated cortical cells [314].

To establish optimal nodule development, two regulatory events, bacterial infection and nodule organogenesis, must be coordinated in the epidermis and cortical cells, respectively, during this process [315]. The symbiotic reactions between the host legume plants and *Rhizobium* are sustained by nodulation factors (Nod factors), which are lipochitin oligosaccharides released by rhizobia [316].

Plant ethylene levels were found to be higher after *Rhizobium* sp. infection of legumes, and this higher ethylene concentration inhibited further rhizobial infection and nodule development [281]. By producing a small compound molecule called "rhizobitoxine", some rhizobial strains can enhance the number of nodules formed on the host bean plant's roots by restricting the rise in ethylene production [317].

Rhizobitoxine is a phytotoxin that inhibits ethylene biosynthesis by chemically inhibiting the enzyme 1-aminocyclopropane-1-carboxylate (ACC) synthase [318]. ACC deaminase is an enzyme produced by some rhizobial strains that eliminates some of the ACC (the immediate precursor to ethylene in plants) before it is converted to ethylene. The plant's nodule production and biomass increase by 25–40% as a result of this reduction [319].

Because around 1–10% of rhizobial strains in the field naturally contain ACC deaminase, it is possible to improve the nodulation effectiveness of rhizobia strains without ACC deaminase by genetically engineering them with rhizobia ACC deaminase genes [268]. The introduction of an ACC deaminase gene from *Rhizobium leguminosarum* bv. *viciae* into the chromosomal DNA of a *Sinorhizobium meliloti* strain that lacked this enzyme enhanced nodule numbers by 35% and host alfalfa plant biomass by 40% compared with the wild-type control strain [295,313].

Azorhizobium is a stem nodule-forming symbiotic bacterium that forms stem nodules and fixes N_2, among other *Rhizobium* strains [320]. They also make a large amount of indole acetic acid (IAA), which helps plants thrive. *Bradyrhizobium* is a good nitrogen fixer, and when it was inoculated into *Mucuna* seeds, it boosted total organic carbon, N_2, P, and K levels in the soil. As a result, it boosted plant growth, soil microbial population, and plant biomass and lowered the weed population [276].

Azospirillum

Azospirillum is a Gram-negative, aerobic nitrogen-fixing bacteria that do not form nodules and belong to the Spirilaceae family [321]. Although there are several species in this genus, such as *Azospirillum amazonense*, *Azospirillum halopraeferens*, and *Azospirillum brasilense*, *Azospirillum lipoferum* and *A. brasilense* are the most beneficial [322].

Because they develop and fix nitrogen on the organic salts of malic and aspartic acid, *Azospirillum* forms associative symbiosis with many plants, notably those with the C_4 dicarboxylic pathway (Hatch–Slack pathway) of photosynthesis [323].

As a result, it is mostly suggested for maize, sugarcane, sorghum, pearl millet, and other crops. They make growth stimulants (IAA, gibberellins, and cytokinin) that help in root development and nutrient uptake (N, P, and K). Inoculation with *Azospirillum* has a significant impact on root development and exudation [324].

When *A. brasilense* sp. 245 was inoculated to maize, the production of various phytohormones increased noticeably, resulting in a significant increase in maize growth [281]. The root physiology and architecture of maize were altered as a result of the increased synthesis of several phytohormones, resulting in an increase in mineral intake by the plant [281]. Inoculation with *Azospirillum* and *Pseudomonas* altered the cultivable bacterial community in the wheat rhizosphere, according to Naiman et al. [325].

They also found that inoculating the soil microflora with *Azospirillum* and *Pseudomonas* altered the profiles of carbon source use during the tillering and grain filling stages [325]. Inoculation with two *A. brasilense* strains (40 and 42 M) isolated from maize roots was also found to affect the community-level physiological profiles of the cultivable microbial communities associated with rice [322].

Azotobacter

Azotobacter is a genus of non-symbiotic, free-living, aerobic, photoautotrophic bacteria belonging to the *Azotobacteriaceae* family. *Azotobacter chroococcum* is the most frequent species in arable soils [326]. They are usually found in neutral and alkaline soils. *Azotobacter vinelandii*, *Azotobacter beijerinckii*, *Azotobacter insignis*, and *Azotobacter macrocytogenes* are among the other species identified [322]. They produce the Vitamin B complex and various phytohormones such as gibberellins, naphthalene acetic acid (NAA), and other compounds that prevent root infections while promoting root growth and mineral uptake [327].

Azotobacter has been found to release chemicals that limit the growth of certain root infections while also improving root growth and nutrient uptake [281]. *Azotobacter* has also been found to add 15–93 kg N/ha to *Paspalum notatum* roots [276]. Another strain, *Azotobacter indicum*, can produce a variety of antifungal antibiotics that are utilized to reduce seedling mortality by inhibiting the growth of many harmful fungi in the root region [328].

Azotobacter populations are often low in the rhizosphere of crop plants and in uncultivated soils, according to research. This organism has been found in the rhizosphere of a variety of crops, including rice, maize, sugarcane, bajra, vegetables, and plantation crops [329].

Blue-Green Algae (Cyanobacteria)

The blue-green algae are photosynthetic organisms that belongs to eight different families. They promote plant growth by generating auxin, indole acetic acid, and gibberllic acid, as well as fixing roughly 20–30 kg N/ha in submerged rice fields [322].

For lowland rice production, nitrogen is one of the main nutrients required in high quantities. Soil nitrogen and biological nitrogen fixation (BNF) by related microorganisms are the two main sources of nitrogen [308,330]. Fungi, liverworts, ferns, and flowering plants create symbiotic relationships with blue-green algae [310]. *Anabena oryzae, Nostoc calcicola,* and *Spirulina* sp. are three blue-green algae that have been shown to reduce the quantity of galls and egg masses induced by the root-knot nematode *Meloidogyne incognita* infecting cowpea, and to improve plant growth [281].

Azolla

Azolla has a 4–5% nitrogen content on a dry basis and 0.2–0.4% on a wet basis. In rice production, it can be a valuable source of organic manure and nitrogen [322]. The important aspect of using *Azolla* as a biofertilizer is that it decomposes quickly in the soil and provides nitrogen to rice plants efficiently. In addition, it adds to the provision of phosphorus, potassium, zinc, iron, molybdenum, and other micronutrients [331].

Prior to rice cultivation, *Azolla* can be utilized as a green biofertilizer in the fields. *Azolla pinnata* is the most commonly used species in India, and it may be produced commercially through vegetative techniques [279]. *Azolla caroliniana, Azolla microphylla, Azolla filiculoides,* and *Azolla mexicana* are some of the other *Azolla* species that have been introduced to India for their huge biomass output [322].

Phosphate Solubilization

Despite the fact that phosphorus is abundant in the soil, the majority of it is insoluble and hence is inaccessible to support plant growth, since plants only absorb it in two soluble forms: monobasic and dibasic. Inorganic phosphorus, such as apatite, or organic phosphorus, such as inositol phosphate (soil phytate), phosphomonoesters, and phosphotriesters, may be present [332].

Furthermore, much of the soluble inorganic phosphorus used in chemical fertilizers is quickly immobilized after being applied to the field. As a result, it is unavailable to plants and hence is wasted [332].

This has prompted researchers to look for environmentally benign and cost-effective ways to boost crop output in low-phosphorus soils. Microbes that can solubilize inorganic phosphorus play a critical role in these settings as a potential option for providing phosphorus to the plants. As a result, they are regarded as a promising biofertilizer, since they may provide the necessary phosphorus to plants, even from low-quality sources [279].

Organic acids with a low molecular weight such as gluconic and citric acids, which are generated by several soil microorganisms, are responsible for inorganic phosphorus solubilization [295]. Figure 14 depicts a schematic diagram of phosphate solubilization by microorganisms. The hydroxyl and carboxyl groups in these low-molecular-weight organic acids can chelate the cations attached to phosphate, resulting in the conversion of insoluble phosphorous to its soluble form.

The mineralization of organic phosphorus, on the other hand, is accomplished by the production of several phosphatases that catalyze the hydrolysis of phosphoric esters [333]. Above all, phosphate solubilization and mineralization can occur in the same bacterial strain [334]. *Pseudomonas, Bacillus, Rhizobium, Burkholderia, Achromobacter, Agrobacterium, Micrococcus, Acetobacter, Flavobacterium,* and *Erwinia* are among the bacteria that have the ability to solubilize insoluble inorganic phosphorus [276].

Phosphate-solubilizing bacteria are commonly found in large numbers in soils and plant rhizospheres. These comprise aerobic and anaerobic strains, with aerobic strains being more common in submerged soils [334]. However, it has been discovered that the rhizosphere has a larger concentration of phosphate-solubilizing bacteria (PSB) than non-rhizosphere soil [276]. PSB stimulate the efficacy of biological nitrogen fixation (BNF) by nitrogen-fixing bacteria, in addition to delivering phosphorus in soluble form to plants [335].

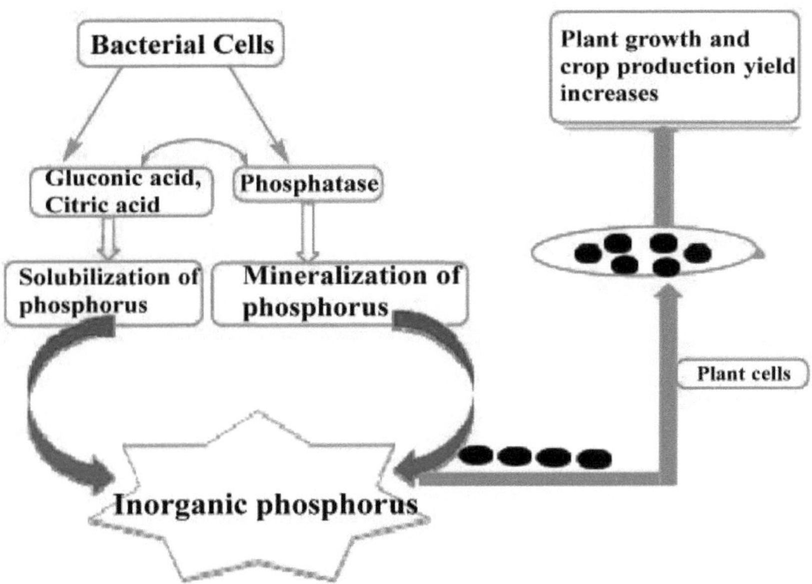

Figure 14

Phosphate-solubilizing rhizobacteria solubilize inorganic phosphorus. Inorganic phosphorus is solubilized by bacteria using organic acids with a low molecular weight such as gluconic and citric acids. These acids' hydroxyl (OH) and carboxyl (COOH) groups chelate the phosphate-bound cations, converting insoluble phosphorus into a soluble organic form. Mineralization of soluble phosphorus is accomplished through the production of several phosphatases, which catalyze the hydrolysis process. When plants absorb these solubilized and mineralized phosphorus molecules, their overall growth and crop output improve dramatically [276].

Iron is an essential ingredient for practically all living things. Iron is required by all plants, animals, and microbes [281]. Iron exists as Fe^{3+} in an aerobic environment and is prone to generating insoluble hydroxides and oxyhydroxides. As a result, the majority of iron is unavailable for absorption by bacteria and plants [281].

In general, bacteria obtain iron via secreting siderophores, which are low-molecular-weight iron chelators with a high affinity for complex iron (Figure 15). The majority of siderophores are water-soluble, and they are classified as extracellular or intracellular siderophores [281]. *Rhizobacteria* differ in their ability to use siderophore cross-linking. Some *Rhizobacteria* use homologous siderophores proficiently, while others use heterologous siderophores [310,336].

Iron is reduced from Fe^{3+} to Fe^{2+} in the bacterial membrane in both Gram-positive and Gram-negative bacteria, and then released into the cell through siderophores via a gating mechanism that connects the inner and outside membranes (Figure 15). Under iron-limiting conditions, siderophores operate as solubilizing agents for iron from minerals or organic molecules [337]. Similar to iron, siderophores create stable complexes with other heavy metals, as well as radioactive particles such as uranium and neptunium [338].

The concentration of soluble metal increases when the siderophores bind to a heavy metal. As a result, bacterial siderophores assist the host plant to reduce the stress caused by elevated heavy metal levels in the soil [295]. Plants absorb iron from bacterial siderophores using a variety of processes, including chelation and release, direct uptake of siderophore–Fe complexes, and ligand exchange reactions [339].

According to Thomine and Lanquar [339], siderophores facilitated iron transfer in oat plants and elevated plant growth. Rhizophore-produced siderophores delivered iron to the oat plant, which possesses a mechanism for utilizing Fe siderophores when iron is scarce [339]. *Pseudomonas fluorescens* C_7 generated the Fe–pyoverdine complex, which was taken up by *Arabidopsis thaliana* plants, resulting in a rise in iron levels in plant tissues and improved plant growth [336].

When plants are exposed to stress situations such as heavy metal pollution, the availability of iron to plants by soil bacteria becomes extremely important. In this case, siderophores can also assist plants to cope with the stress caused by high amounts of heavy metals [338].

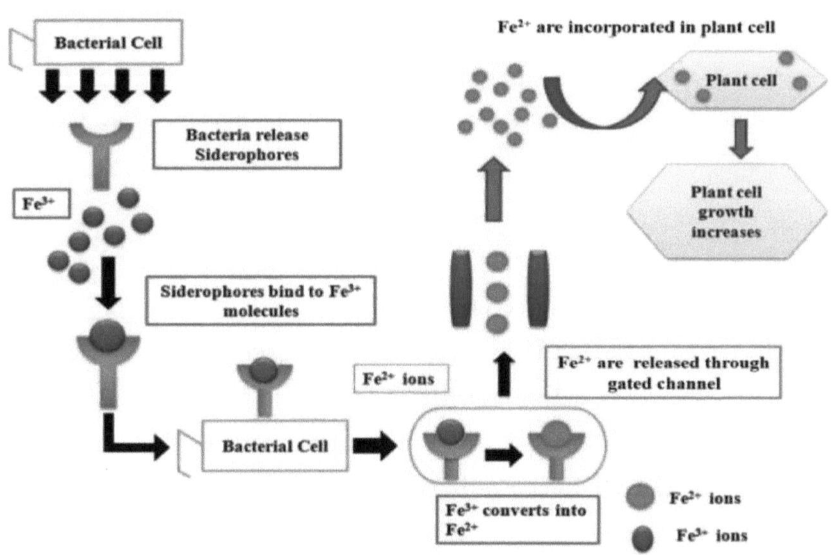

Figure 15

Plant growth-promoting rhizobacteria produce siderophores, which are used to sequester iron. Bacteria release low-molecular-weight iron chelators known as "siderophores," which have high affinity for Fe^{3+}, bind firmly to it, and are taken up by bacteria. Fe^{3+} is converted to Fe^{2+} inside the bacterial membrane, and Fe^{2+} is discharged into the cell via a gated channel that connects the bacteria's inner and outer membranes. The total plant growth improves significantly when the host plant integrates these soluble Fe^{2+} molecules produced by the bacteria.

Modulation of Phytohormone Levels

Plant hormones, also known as phytohormones, play a crucial role in plant growth and development [281]. When plants are exposed to growth-limiting environmental conditions, evidence suggests that they modify their endogenous phytohormone levels to reduce the detrimental impacts of environmental stress [281].

Microorganisms in the rhizosphere have been found to produce or modify phytohormone levels in the host plants. As a result, by modifying the level of endogenous phytohormones in the host plants, they can considerably influence the hormonal balance and stress response of the host plant [295].

For a long period, scientists have known that bacteria produce the phytohormone auxin (indole-3-acetic acid/indole acetic acid/IAA). According to one study, 80% of microorganisms isolated from the rhizosphere of diverse crops are capable of generating and releasing auxins as secondary metabolites [337].

Indole-3-acetic acid (IAA) is involved in many aspects of plant growth and development, as well as defense responses. The exceptional complexity of IAA biosynthesis, its transport mechanisms, and the different signaling pathways involved in IAA synthesis and transport reflects this diversity of roles [340].

In general, IAA stimulates seed and tuber germination; increases the rate of xylem and root development; controls vegetative growth processes; initiates lateral and adventitious root formation; mediates responses to light, gravity, and fluorescence; and affects photosynthesis and pigment formation, the biosynthesis of various metabolites, and stress resistance [293].

Because IAA is involved in various cell division and vascular bundle creation processes, it appears that a higher level of IAA in the host legume plants is required for nodule development [294]. Furthermore, bacterial IAA increases the root surface area and root length, allowing the plant to acquire soil nutrients more easily [294].

Furthermore, rhizobacterial IAA loosens plant cell walls, allowing for greater root exudation, which offers additional nutrients to sustain bacterial development [294]. As a result, rhizobacterial IAA has been identified as a crucial effector molecule in both disease and phytostimulation in plant–microbe interactions [294].

The amino acid tryptophan is an important component that influences IAA production levels. Tryptophan has been identified as the principal precursor of IAA and has been shown to have an important role in altering IAA biosynthesis levels [281]. Starting with tryptophan, at least five distinct processes for the synthesis of IAA have been reported, most of which are comparable with the mechanisms discovered in plants, but a few intermediates differ in each case [341].

The production of IAA via indole-3-pyruvic acid and indole-3-acetic aldehyde is the first pathway. The majority of bacteria, including *Rhizobium*, *Azospirillum*, *Erwinia herbicola*, *Klebsiella*, and others, use this pathway. The conversion of tryptophan to indole-3-acetic aldehyde is the second process, which may include an alternate pathway by which tryptamine is generated. *Pseudomonas* and *Azospirilla* use this route. The biosynthesis of IAA occurs via indole-3-acetamide in the third route. *Agrobacterium tumefaciens,*

Pseudomonas syringae, and other phytopathogenic bacteria use this pathway. The conversion of tryptophan into indole-3-acetonitrile is the fourth step for IAA biosynthesis. Cyanobacteria have this mechanism. The last mechanism, which is more widespread in plants, Cyanobacteria, and *Azospirilla,* is the production of IAA via a tryptophan-independent pathway.

Although bacterial IAA has been implicated in almost every aspect of plant growth and development, the acquisition of bacterial IAA may modify the endogenous pool of plant IAA. The degree of endogenous IAA in plants is critical in determining whether bacterial IAA stimulates or hinders plant growth in this respect. Endogenous IAA has been determined to be either ideal or sub-optimal for plant root development [295].

Synthesizing multiple types of antibiotics is the most common way for plant growth-promoting bacteria (PGPB) to limit plant pathogen proliferation [342,343]. Many of the compounds have been thoroughly researched and some have even been marketed. The majority of commercialized rhizobacterial products function as bio-inoculants to combat plant diseases rather than to improve plant nutrition or reduce abiotic stressors [342].

Plant illnesses caused by pathogens such as *Fusarium* spp., *Pythium* spp., *Rhizoctonia* spp., and *Sclerotium* spp. have been reported to be treated by using biofertilizers such as *Trichoderma harzianum, P. fluoresecens,* and *Bacillus subtilis,* which boost plant growth and overall output. Hydrogen cyanide (HCN), phenazines, pyrrolnitrin, 2,4-diacetylphloroglucinol, pyoluteorin, viscosinamide, and tensin are among the antifungal metabolites produced by various *Rhizobacteria* [303].

It has also been observed that the contact between some *Rhizobacteria* and plant roots can protect the host plant from pathogenic fungi, bacteria, and viruses. Induced systemic resistance (ISR) is the term for this phenomenon [344]. Furthermore, ISR does not necessitate any direct interaction between the pathogens and the resistance-inducing PGPB [295].

Induced systemic resistance (ISR) is caused by jasmonate and ethylene signaling in the host plant, which acts as a defense mechanism against a range of plant pathogens [295]. Many individual bacterial components, such as lipopolysaccharides (LPS), flagella, siderophores, cyclic lipopeptides, 2,4-diacetylphloroglucinol, and homoserine lactones, as well as volatile compounds such as 2,3 butanediol and acetonin, have been reported to cause ISR in the host plant, allowing the host plant to combat a variety of plant pathogens [344].

Some biocontrol bacteria generate enzymes such as chitinases, cellulases, 1,3-gluconases, proteases, and lipases that can lyse a section of the cell wall of many pathogenic fungi such as *Botrytis cinerea, Sclerotium rolfsii, Fusarium oxysporum, Phytophthora* spp., *Rhizoctonia solani,* and *Pythium ultimum* [345,346].

Some PGPB strains produce siderophores, which operate as a biocontrol agent. In this approach, PGPB's siderophores prevent pathogens from acquiring adequate iron, limiting their growth and proliferation [345]. Because the siderophores produced by PGPB have a higher affinity for iron than the pathogens, this technique is effective. As a result, the infections' ability to utilize iron is diminished and they are unable to multiply in the rhizosphere [295].

Plants have been reported to synthesize ethylene in response to a range of stressors, including fungal phytopathogenic infections [347]. When plant cells become infected, ethylene causes a stress/senescence response in the plant, which results in the death of cells that are either infected or present near the fungal infection site [347].

As a result, increasing levels of ethylene build up, as well as the infection caused by plant pathogens, causing a large amount of the harm to the plant. Exogenous ethylene has also been shown to exacerbate the severity of fungal infections. As a result, lowering the ethylene response is one strategy to reduce the harm produced by phytopathogen infections of the host plants [268].

Ethylene inhibitors have been found to not only reduce the ethylene response level but also to diminish the severity of fungal infections. When the host plant is affected by pathogens, the enzyme ACC deaminase found in PGPB can adjust the ethylene level [281].

As a result, the most straightforward strategy to reduce ethylene levels is to apply PGPB harboring the ACC deaminase gene to the plants (usually the roots or seeds).

Benefits of Biofertilizers in Food Production

To meet the increased need for food, continuous and indiscriminate usage of synthetic or chemical fertilizers has unquestionably resulted in contamination and ecosystem modification [281]. Even so, the long-term impacts of using synthetic or chemical fertilizers lower soil fertility and have resulted in the production of disease-prone crops [348,349].

The amount of food produced today compared with the amount required to feed everyone in 2050 is drastically lower. By 2050, the world's population will have swelled to about 10 billion people, with roughly 4.5 billion more mouths to feed than in 2022.

People will consume more resource-intensive, animal-based diets as their wages rise. To feed the growing population with a deficit amount of available nutrients, the world certainly needs to encourage agricultural productivity in a sustainable and ecofriendly way. Hence, it is necessary to re-evaluate many of the existing agricultural approaches, which include the use of chemical fertilizers, pesticides, herbicides, fungicides, and insecticides [350].

In light of the harmful effects of chemical or synthetic fertilizers, biofertilizers are supposed may be a safe alternative to chemical inputs and minimize alteration of the ecosystem to a great extent. Biofertilizers are cost-effective and ecofriendly in nature, and their prolonged use enhances soil fertility substantially [281].

It has been found that using biofertilizers increases crop yield by 10–40% by increasing protein, vital amino acids, and vitamins, and nitrogen fixation [351]. Biofertilizers provide a number of advantages, including being a low-cost source of nutrients, excellent suppliers of micro-compounds and micronutrients, organic matter suppliers, growth hormone producers, and a means of counteracting the negative effects of chemical fertilizers [352].

Different microorganisms are important components of soil, and they play a key role in a variety of biotic activities in the soil ecosystem that keep the soil active for nutrient mobilization and long-term crop development [310].

Will the Improvement of The Technical Environment Help Promote Organic Fertilizer and Bio pesticides Farmers?

Substituting organic fertilizers for chemical fertilizers is an effective measure to curb the degradation of the agricultural ecological environment, stabilize agricultural production, and promote sustainable agricultural development [353,354]. Excessive application of chemical fertilizers is widespread in China's agricultural production [355].

However, the negative effects caused by the irrational application of chemical fertilizers have gradually emerged. As a result, the agricultural non-point source pollution problem brought about by the excessive input of chemical inputs has gradually evolved into an environmental problem [356- 358].

Chemical fertilizers have played an essential role in the development of agricultural production in China; compared with 2000, the average unit area output of China's three major staple grains in 2020 increased by 31.2%, and the increase in vegetable production was incredible. However, with the continuous improvement of dependence on chemical fertilizers, problems such as unreasonable fertilization intensity and fertilization structure in agricultural production have become increasingly prominent.

As a result, the amount of chemical fertilizer application per unit area in China's agricultural production has increased from 273.2 kg/hm^2 in 2001 to 313.50 kg/hm^2 in 2020, reaching the highest value in history in 2014. This rate is above the global average. Due to rapid economic and social development, urban and rural residents' consumption levels and social structures have undergone tremendous changes.

As a result, the requirements for the quality and safety of agricultural products have been continuously improved, especially since food consumption is no longer limited to the level of food and clothing. The rapid demand for green agricultural products has become an inevitable trend.

The changes in demand for agricultural products will inevitably force the transformation of agricultural production to new themes, such as sustainable and organic. In view of the above problems, the Chinese government has established the reduction of chemical fertilizers as a phased task of agricultural production. In 2015, the government put forward the goal of "zero growth of chemical fertilizers by 2020".

Since 2016, Central Document No. 1 has continuously established the reduction of agricultural fertilizers as an important work goal. In 2017, the government has further promoted green agricultural development, including the use of organic fertilizers instead of chemical fertilizers. The target of this work goal is to change the current unreasonable fertilization structure and fertilization methods, and to achieve partial substitution of chemical fertilizers through the use of new agricultural production technologies. It helps to achieve the ultimate goal of chemical fertilizer reduction and agricultural green development.

Farmers are the micro-subjects of agricultural production and the behavioral decision-making units that respond to the push towards the reduction of chemical fertilizers and the green development of agriculture [359].

Thus, exploring the relationship between the production behavior of farmers and their technical efficiency is a necessary way to promote and coordinate the dual goals of green agricultural development and ensure farmers' income increases. Technical efficiency is the gap between farmers' production and technological frontiers and is a visual embodiment of farmers' production and management capabilities. The excessive application of chemical fertilizers will inevitably reduce its marginal output [360].

Applying organic fertilizers instead of chemical fertilizers can reduce excessive fertilizer inputs to reduce costs and improve the marginal output of organic fertilizers. In addition, organic fertilizers can improve the soil environment and quality and further optimize the marginal output of other production factors [361]. Therefore, using organic fertilizers instead of chemical fertilizers may have significant advantages for improving farmers' technical efficiency and may also make significant contributions to the green development of agriculture.

Vegetables are the largest cash crops in China, and their sown area reached 21.485 million hectares in 2020, making them the second largest crop after food crops. For vegetables, fertilizer input is related to environmental impact. At the same time, product quality and safety are closely related, so this study selected vegetable farmers as the research object.

The use of organic fertilizers by farmers is a sustainable production behavior. This study first explores the literature from the perspective of farmers' green production behavior and then reviews the relevant research on substituting chemical fertilizers with farmers' organic fertilizers.

Research on green agricultural production behavior is generally carried out from the perspective of willingness and behavior. From the perspective of willingness to adopt, farmers have different performances across various green production technologies, production factors, or production methods. The difficulty and cost of these production technologies, factors, or methods are essential factors affecting farmers' willingness [362,363]. From the perspective of adoption behavior, the proportion of farmers in China adopting green production behavior is low [364,365].

Some scholars considered the correlation between different agricultural production technologies; they regarded green agricultural production behavior as a combination of agricultural production technologies, and came to similar conclusions [366,367]. The influencing factors of adoption behavior are more complex and diverse than those of farmers' willingness to adopt. They include farmer endowments [368], social networks [369,370], land characteristics [371,372], government subsidies [373,374], and other factors.

Many studies have focused on the effects of green production behavior on farmers. Some scholars believe that green production behavior has positive effects. For example, some studies have found that adopting integrated bio-textile technologies can reduce environmental costs and total costs, thereby increasing total returns [375]. Adopting arable land protection technology can increase the rice yield of farmers [376]. The adoption of green production technology can significantly improve agricultural production efficiency [377].

Some scholars also believe that green production behavior has no significant effect. The use of conservation tillage techniques has not effectively increased the yield of wheat and maize in the Yellow River Basin, which may be related to the scale of farmers' operations and the proportion of farmers' adoption in the surrounding areas. Some scholars believe that the effect of green production behavior will also be affected by several external conditions [378,379].

Research on adopting organic fertilizers is also prevalent, and many scholars regard organic fertilizers as a green factor of production and include them in their research frameworks [380,381]. Since organic fertilizer is an input-based production technology, it can improve soil structure, environmental quality, and fertility [382].

The stability of land rights has been the focus of such studies. Most scholars believe that stable land property rights can promote the adoption of organic fertilizers [383].

On the other hand, some scholars believe that the stability of land rights will not affect farmers' use of organic fertilizers [384]. In addition, adopting organic fertilizer as a green production technology is also affected by a series of factors, including farmer characteristics, land endowments, and policy subsidies. Therefore, organic fertilizer as a production technology will not exist independently and be widely adopted. Many scholars have studied such problems by considering the interaction effect of organic fertilizer with other production factors or production technologies [385,386].

Studies on the effect of replacing chemical fertilizers with organic fertilizers are more specific, and the conclusions are more consistent. Replacing chemical fertilizers with organic fertilizers can significantly improve soil quality, which is beneficial to crop drought, pest presence, and lodging resistance. In addition, the yield per mu of the fertilized plots was higher than that of the unfertilized plots [387,388].

In recent years, the number of studies exploring the application effect of organic fertilizers from the economic perspective has gradually increased. Based on microdata studies in Hubei and Hunan, some studies found that using organic fertilizers can significantly reduce agricultural non-point source pollution and domestic waste [389]. Some studies suggest that when compared with chemical fertilizers, organic fertilizers can improve production efficiency more significantly [390].

Some studies found that using organic fertilizers can increase farmer incomes by 2661~2959 ETB in Ethiopia (ETB refers to the currency code of Ethiopian Birr.) [**391**].

In summary, professors and scholars in China and abroad have conducted many analyses of farmers' green production behavior, organic fertilizer adoption behavior, and corresponding effects from different perspectives. As a result, multidisciplinary research has dramatically enriched the academic perspective of substituting chemical fertilizer with organic fertilizer. However, there are still the following aspects to be deepened and improved.

First, the relevant research on the use of organic fertilizers is still primarily found in the field of natural sciences. However, this study aims to conduct its analysis from the economic perspective to assess the impact of organic fertilizer substitution on farmers' production. Second, studies on the effect of organic fertilizer substitution of chemical fertilizers primarily focus on the impact on farmers' income and yield.

In contrast, this study focuses on the impact of organic fertilizer substitution on the technical efficiency among farmers. Third, studies are primarily based on micro-survey data, but there is a non-random behavior of farmers using organic fertilizers. Therefore, the problem of "self-selection" often occurs.

Theoretical Mechanisms and Research Methods

Influence Mechanism of Organic Fertilizer Substitution of Chemical Fertilizer on the Farmers' Technical Efficiency

Technical efficiency is the output efficiency brought about by technological progress or the improvement of management capabilities. It reflects the maximum output capacity of farmers at a given input level or the ability to achieve the optimal allocation of input factors when achieving a specific output. Therefore, from the perspective of input and output, this study analyzes the changes brought about by substituting organic fertilizers for chemical fertilizers on the input and output of farmers' production and then clarifies the impact mechanism on farmers' technical efficiency (**Figure 16**).

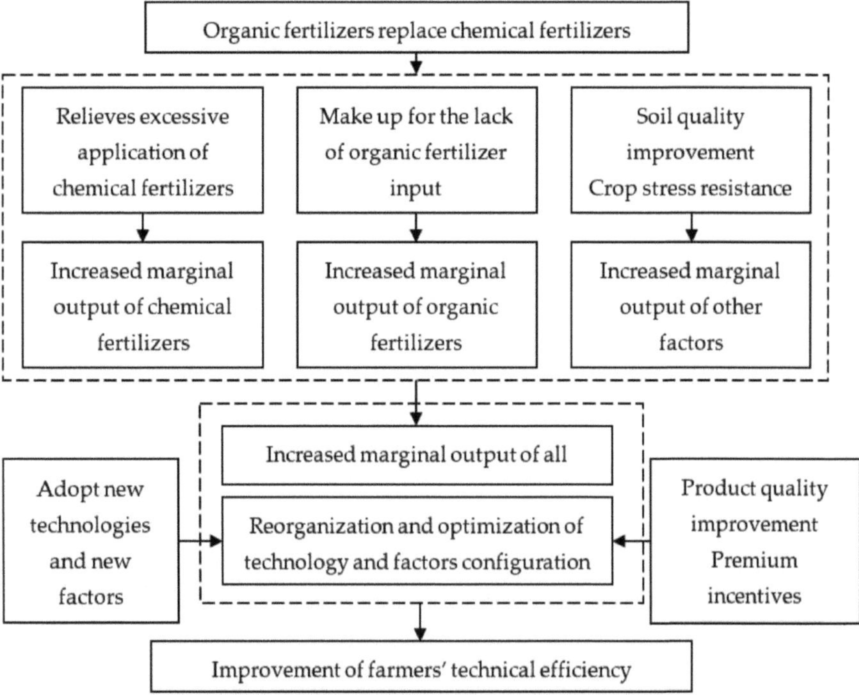

Figure 16. Influence mechanism of organic fertilizer substitution of chemical fertilizer on the farmers' technical efficiency.

First, the marginal output of fertilizer inputs can increase. Excessive input of chemical fertilizer is common. This has adverse effects on the agricultural production environment and causes waste of resources, increases unnecessary costs, and reduces the marginal output of chemical fertilizer. However, organic fertilizer input in agricultural production is still insufficient [392].

Therefore, if organic fertilizer is used to replace some chemical fertilizers, the marginal output of both organic and chemical fertilizers will be significantly improved. Therefore, substituting organic fertilizers for chemical fertilizers will significantly increase the marginal output of fertilizers as an input factor, thereby improving farmers' technical efficiency.

Second, the marginal output of other inputs can increase. Numerous studies have shown that using organic fertilizers can effectively improve soil quality while improving the resilience of crops to stress [393].

This improvement in soil quality and crop resilience means that producing the same agricultural products on the same land requires less input from machinery, labor, and other factors. From another point of view, this is the increase in the marginal output of other factors such as machinery and labor. Therefore, replacing chemical fertilizers with organic fertilizers can improve the quality of some input factors in agricultural production to improve the marginal output of other factors, thereby promoting the improvement of farmers' technical efficiency.

Third, the optimization of the technology portfolio is caused by substituting production technology, elements, and complementarity. Farmers' adopting new production technology is a systematic behavior [394]. When farmers adopt new technology, they may use some other technologies simultaneously to achieve technological complementarity or abandon part of the technology to carry out technological substitution to maximize their interests. These behaviors will promote the further optimization of their technology portfolio.

At the same time, production factors have complementary or substitution effects. As a green and environmentally friendly agricultural production technology, the adoption of organic fertilizer will inevitably lead to the reorganization and optimization of farmers' production technology and factor allocation.

Fourth, the quality of agricultural products is improved, and premium incentives are encouraged. According to existing studies, using organic fertilizer can effectively improve the quality of agricultural products, thus leading to a price premium compared to ordinary agricultural products. The higher the quality of agricultural products, the higher the premium level obtained [395].

For small farmers, the adoption of new technologies or new inputs will lead to higher costs and risks, and this premium will provide incentives for farmers to adopt the new technology or input. New technologies further optimize resource allocation, encourage farmers to upgrade their technology, and optimize resource allocation, forming a virtuous circle that continuously improves their technical efficiency.

In summary, farmers using organic fertilizers in place of chemical fertilizers can improve their technical efficiency by strengthening the marginal output of production factors and promoting the restructuring and optimization of production technology and factor allocation. At the same time, it also provides long-term incentives for this promotion through price premiums. Based on the above analysis, the hypothesis proposed by this study is that farmers' using organic fertilizer to substitute chemical fertilizers can improve technical efficiency.

There are a variety of indicators to measure agricultural productivity, including but not limited to productivity, technological efficiency, and total factor productivity. Among them, single-factor productivity, such as land and labor productivity, measures agricultural production efficiency through the ratio of individual factors to agricultural output.

Total factor productivity is the part of output growth that deducts the increase in factor inputs. In actual research, the growth of total factor productivity can be deconstructed into the rate of technological progress, return on the scale, improvement of technological efficiency, and improvement of allocation efficiency. Among them, technical efficiency refers to the degree to which farmers master and utilize a particular technology, which is the ratio of actual output to the boundary production function; allocation efficiency is the adjustment of input and output relative to the price after the production technology is selected [396].

In this study, organic fertilizer substitution was considered a production technology and its effects were analyzed; as such, technical efficiency was selected as the core explanatory variable. In the current study, the measurement of technical efficiency is primarily conducted using Stochastic Frontier Analysis (SFA) and Data Envelopment Analysis (DEA).

On the one hand, the advantage of the DEA method is that there is no need to set the specific function form to avoid the structural deviation caused by the mis-setting of production functions, such as traditional accounting methods and SFA methods. On the other hand, undesired outputs can be integrated into a unified input and output production system [397].

However, the DEA method also has shortcomings. Although it can measure the efficiency problem of multi-input and multi-output, it ignores the influence of random error [**398**]. In other words, in the actual analysis, the resulting bias may be caused by the existence of random error., It is difficult to test the overall significance of the regression results using the DEA method, thus it is impossible to directly analyze the influencing factors of efficiency [**399,400**].

The SFA method is superior to DEA in that it considers the impact of random error on the results, which determines the production function form In advance. As a result, it can improve the accuracy of calculation technology efficiency and analyze the correlation between efficiency and influencing factors.

The factor influencing technical efficiency is a complex system, so a series of control variables need to be added to the model. Previous studies have focused primarily on the personal characteristics and family characteristics of the head of the farmer's household, including age, education level, identity, part-time level, and the number of laborers [**403**].

This study aims to explore the substitution effects of organic fertilizer, so the farmers' awareness of the quality and safety of agricultural products is also added. Farmers' land resource endowments are also of concern, such as land scale, land fragmentation degree, and topographic conditions [**404,405**].

In addition, the peculiarities of vegetable production, planting density, production experience, and training participation have also been added. The fragility of agricultural production processes has led to disaster situations and participation insurance.

To achieve an effective match between farmers who use organic fertilizers instead of chemical fertilizers and farmers who do not, it is first necessary to conduct regression analysis on the conditional probability fair values of farmers using organic fertilizers instead of chemical fertilizers. To be specific: (1) characteristics of the household head: age has a significantly negative influence, indicating that the older the age, the lower the ability to accept new production factors and production technologies. (2) Family characteristics: the level of part-time employment shows a significant adverse effect because full-time farmers can obtain higher incomes and provide sufficient financial support for adopting new technologies and new factors. (3) Land characteristics: the scale of operation has a significant negative effect; this is because the more extensive the scale of a farmer's operation, the lower the probability of farmers using organic fertilizers will be, in line with the expansion of scale in order to reduce production costs. (4) Production characteristics: the production experience significantly reduces fertilizer substitution. This may be due to the relatively fixed production mode of small farmers. With the increase in production years, their decisions become more "empiricist" and they are inclined to use traditional production factors and technologies. Participation in insurance can significantly enhance fertilizer substitution. Because participation in insurance can bring stable expected returns to farmers, this factor enhances farmers' ability to resist risks and provide a bottom-up guarantee for farmers to adopt new technologies and elements. Training can significantly improve the probability of farmers replacing chemical fertilizers with organic fertilizers because training can give farmers a more objective and accurate understanding of the advantages and disadvantages of chemical fertilizers and organic fertilizers. The disaster situation is significantly negatively correlated with the farmer's fertilizer substitution. When encountering natural disasters, the farmer adds more fertilizer input in the short term to increase the yield effect and make up for the loss. (5) Regional characteristics; there are significant differences in the substitution of organic fertilizers by farmers in central vegetable-producing counties and non-primary-producing counties.

Two primary conditions must be met to ensure the effectiveness of the propensity score matching method. One is the typical support hypothesis. The preference scores of farmers who use organic fertilizers instead of chemical fertilizers and farmers who do not use organic fertilizers instead of chemical fertilizers must have a large common support domain. Plotting the kernel density function plot of the propensity score was used to test the first condition. Taking nuclear matching as an example, it can be seen in **Figure 17** that there are apparent differences in the probability distribution of the propensity scores of the two groups of samples before and after the matching. In contrast, the probability distribution of the propensity scores of the two groups of samples after the match is closer, the matching effect is better, and the expected support hypothesis is met.

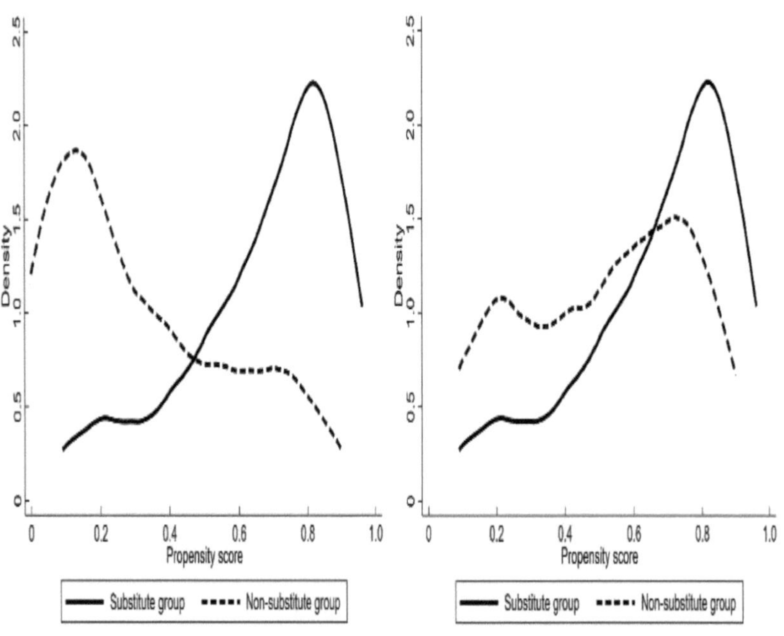

Figure 17. The co-support domains of substitute group and non-substitution group before matching (**left**) and after matching (**right**).

Reducing Environmental Contamination of Fertilizers and Increasing its Efficiency

One of the world's biggest and most impressive studies shows us that simple interventions can produce large results. In a decade-long trial, researchers worked with 21 million smallholder farmers across China to see if they could increase crop yields while also reducing the environmental impacts of farming.[407] They were successful.

In the decade from 2005 to 2015, average yields of maize, rice and wheat increased by around 11%. At the same time, nitrogen fertilizer use *decreased* by around one-sixth. By producing more crops and needing less fertilizer, this experiment provided an economic return of US$12.2 billion. This wasn't achieved through major technological innovations or policy changes: it involved educating and training farmers on good management practices.It's often assumed that fertilizer use – alongside the pollution it creates – and crop yields present an inevitable trade-off. To increase yields, you need more and more fertilizer. This large-scale study suggests this trade-off is not always as extreme as we might think.

To be clear: fertilizers are vital for global food production. There are few innovations that have transformed the world as much as synthetic nitrogen.

For most of human history, food production was limited by the amount of reactive nutrients that were available for crops. This all changed with Fritz Haber and Carl Bosch. Rather than relying on the scarce nitrogen that exists naturally within the world's soils, we could produce our own. Their innovation (the Haber-Bosch process) at the beginning of the 20th century enabled the lives of billions of people.[408] Estimates suggest that every second person reading this has them to thank for being alive today.

Fertilizers help us to achieve higher crop yields. This is an obvious net positive for humans: farmers can produce and earn more, and the world has more food. What's less obvious is that this has a large environmental benefit. Higher crop yields mean we need to use less land for farming.[409] This means we can protect forests and maintain natural habitats.

But it's true that alongside the environmental benefits, there are also some downsides. Not all of the nitrogen we use is used by the crops. The rest runs off the soils and into the natural environment: fertilizing the rivers and lakes and thereby upsetting the balance of ecosystems and causing biodiversity loss.

We might assume that there is nothing we can do: that to achieve higher yields we need more inputs and therefore necessarily cause more pollution. In this article I show that farmers in many countries *can* reduce fertilizer use without sacrificing food production.

There are large differences in fertilizer use across the world

Crops, like any organism, need nutrients to grow. When particular nutrients are lacking, they fail or grow at a much slower rate. These are called 'limiting nutrients'. What nutrient is the most limiting varies across the world: some soils lack nitrogen, while others lack phosphorus or potassium.

If a soil is lacking nutrients naturally we can add our own. This can be in the form of synthetic fertilizers, or organic additions such as manure. There are very large differences in how much fertilizer is applied across the world. We see this in the charts below: first as a map of average fertilizer use per hectare of cropland; and second, with the breakdown by nutrient in the bar chart.

There are 100-fold differences between countries. In many of the world's poorest countries – particularly across Sub-Saharan Africa – farmers apply only a few kilograms of fertilizer per hectare. For context, one hectare is about 1.5-times the size of a football pitch.[410] Contrast this with countries such as China, Brazil, the UK or Egypt, where farmers apply hundreds of kilograms per year. They apply as much in a few days as some farmers do in an entire year.

This has led to a divided world:

- In many poorer countries we need *more* fertilizers. Improvements in crop yields have been slow, and large yield gaps could be closed through more and better management of inputs.[411] This is not only good for farmers, but also for the environment: for the reasons above, closing yield gaps is one of the best ways we can prevent habitat loss across the tropics.

This is why it's damaging for agencies, such as the United Nations Development Programme to continually promote the message that the less fertilizer, the better. It's not good for humans, or the environment.

- But, as we will see, many countries are *overapplying* nitrogen. They could cut back without negatively affecting their crop yields.

Nitrogen use efficiency: balancing yields and the need for nutrient inputs

Using lots of fertilizer wouldn't necessarily be a bad thing if all of it was used by the crops. Unfortunately, most of it isn't.

To capture this, we can look at the ratio of nitrogen in harvested products (our crops) compared to our inputs (fertilizers or manure); this ratio is called the 'nitrogen use efficiency' (NUE). A NUE of 60% would mean that the amount of nitrogen in our crops was 60% of the nitrogen that was added to them as inputs. The remaining 40% of nitrogen was *not* used by the crops.

A low NUE is bad. This means very little of the nitrogen we add is taken up by the crops. A NUE of 20% would mean that 80% of the applied nitrogen became a pollutant.

Soon we will see that some countries have a *very* high NUE – greater than 100%. You might assume that this is good news. In fact, it's often the opposite. This means they are *undersupplying* nitrogen, but continue to try to grow more and more crops. Instead of utilizing readily available nutrients, crops have to take nitrogen from the soil – a process called 'nitrogen mining'. Over time this depletes soils of their nutrients which will be bad for crop production in the long-run.

Globally, NUE has been stubbornly low, at 40% to 50% since 1980.[412]This is surprisingly low. It means that less than half of the nitrogen we apply to our crops is actually taken up by them. The rest is excess that leaks into the natural environment.

But there are very large differences in NUE across the world, as shown in the map. Some countries achieve low NUE – less than 40%. Both India and China, for example, have an efficiency of only one-third. Some countries, though, do much better. France, Ireland, the UK, and the US, have an efficiency greater than two-thirds.

How nitrogen use efficiency has changed over time

We can reduce nitrogen pollution without a decline in yields

So, nitrogen efficiency rather than just fertilizer use seems like a better sustainably metric for us to benchmark.

We might assume that all countries could achieve the same high NUE. But, maybe it's still unfair to compare countries across the world in this way. Differences in climate, vegetation, and soil types mean we can't achieve the same yields with the same inputs everywhere. Some countries might have more favorable environmental conditions than others.

How can we better understand which countries are doing well in these yield-fertilizer trade-offs?

An interesting way to tackle this question is to look at the discontinuities of yields and nitrogen pollution at international borders. This is the approach that David Wuepper and his colleagues took in a recent study, published in *Nature*.[413] By looking at the discontinuities of yields, nitrogen balances and inputs across borders the researchers investigated the role that each country's agricultural policies play.   This is because the environmental conditions, climate and soil qualities should be very similar just across the border. Technically they *should* be able to achieve a similar level of NUE, and similar yields. If there are large differences in yields or pollution between one country and its neighbor, we would therefore assume there are important country-specific effects playing a role. It mimics a 'natural experiment' where the environmental conditions are held constant, and policy decisions are the changeable variable.

The contrast at the border between Kazakhstan and China; and Turkey and Syria provide good examples of this. We can see this in the aerial shots. The conditions for growing crops on either side should be similar. But China and Turkey have much more vegetation than their neighbors as a result of nutrient inputs and how they manage agriculture.

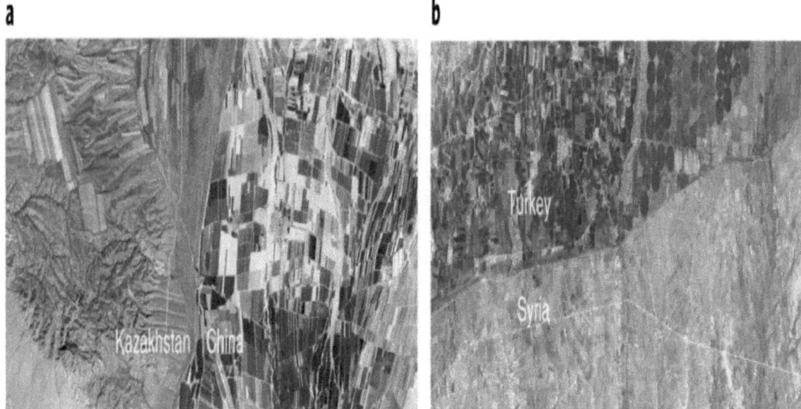

Discontinuities in vegetation at country borders[414]

Using satellite imagery and geospatial datasets these researchers could measure four key metrics at high-resolution across hundreds of thousands of cross-country borders: cropland nitrogen balances, nitrogen pollution, yield gaps (the amount that yields could be increased with better management of nutrients), and the natural vegetation potential. They found cross-border differences in the first three metrics, but not in natural vegetation potential. This is important because it means our assumption that the environmental conditions on either side of borders is similar, is a valid one.[415]

Across this large global dataset, the researchers found that the discontinuity in nitrogen pollution across borders was much larger than the discontinuity in yield gaps. Their results suggest that globally there is massive potential to reduce nitrogen pollution without impacting crop yields.

They conclude that nitrogen pollution could be reduced by around 35% if polluting countries became as efficient as their neighbors. This would have little impact on crop yields – increasing yield gaps by only 1%.

Their results also allow us to understand what countries are using nitrogen inefficiently. The map here shows how countries compare in levels of nitrogen pollution versus their yield gains *relative to their neighbors*. Positive values – shown in orange and red – mean a country causes *more* pollution than necessary for the yields that it achieves. Negative values – shown in blue – means a country causes *less*.

There are a couple of important points we need to keep in mind. All of these values are measured *relative to a country's neighbors*. A country might have a good score because their neighbor gets very low yields: South Korea is a good example. Or a country scores well because its neighbor uses nitrogen inefficiently: Mongolia is a good example.

China has the highest score of 170%. This means it causes 170% more nitrogen pollution than is necessary to achieve its level of crop yields. Brazil, Mexico, Colombia, and Thailand also create a lot of pollution. These are the countries that are overapplying nitrogen the most: they could probably reduce fertilizer use significantly without affecting their crop yields.

We often assume that more pollution is an unavoidable cost of trying to close yield gaps. But this trade-off does not always exist.

How can we use nitrogen more efficiently?

You might notice that most of the largest polluters are middle-income countries. During the 1960s and 1970s, many of today's middle-income countries kickstarted their 'Green Revolution' and achieved large increases in food production. Governments offered subsidies for farmers to use fertilizers and other inputs. This made fertilizers cheap and reduced the incentives for farmers to use it efficiently.[416] This cheap fertilizer is one of the reasons that these countries massively overapply nitrogen today.

One way that governments can therefore reduce nitrogen pollution is to adjust the ratio of fertilizer prices to the return on agricultural products. They can adjust subsidies to make it costly for farmers to overuse fertilizers. Instead, they could re-allocate these financial resources towards practices that have positive environmental impacts.

Another option is to invert the financial incentives: rather than subsidizing fertilizers, you could tax them.

We might want to make fertilizers more expensive for countries that overuse them. But we actually want to do the opposite for countries with large yield gaps. As we saw earlier, many countries across Sub-Saharan Africa use barely any fertilizer at all. They achieve very poor yields as a result. Providing subsidies for fertilizers and other inputs would be of massive benefit.

One of the challenges of putting fertilizer on your crops is that it can be hard to know where it is needed. Some parts of your field might be lacking in nitrogen while others have more than enough. Often the easiest and quickest solution is to apply it everywhere, especially if fertilizers are heavily subsidized and cheap. But with emerging technologies, we can do better. Thanks to information from drones or satellite imagery, we can implement 'precision farming', which allows us to see exactly where fertilizers are needed the most.[417] Plant breeding technologies could also offer new opportunities.[418]We can try to improve how efficient we are at using nitrogen, but there's an opportunity to improve how efficiently plants use it too.

Let's not forget that one of the most promising solutions – and one we often overlook – is the simplest and oldest of all. Legumes – crops such as beans, peas and lentils – perform their own magic when it comes to nitrogen. They have the ability to capture nitrogen in the atmosphere and transform it into reactive nitrogen on their own. This is called 'biological fixation'. Unlike most other crops where we have to add additional nitrogen, they create it by themselves. Growing more legumes – either on their own, or alongside other crops – is one of the easiest ways that we can bring nitrogen into the soil.

Finally, there's a lot that we can do by training farmers to adopt sustainable management practices. The 21-million-farmer study in China makes this clear. Large policy changes and technological advancements are often needed to make a large difference, but we shouldn't underestimate the impact that education can make.

Many view crop yields and environmental pollution as an unavoidable trade-off. It doesn't have to be. We can reduce pollution a lot without reducing crop yields. Less pollution, more food, higher farmer returns, and less farmland make this a problem with multiple wins if we can implement the right solutions.

BOMFs play a crucial role in stabilizing the needs of intense agriculture and its impact on soil health and ecosystem functions driven by the resident microbial community. The amalgamation and application of organic matter and low-grade mineral sources of potassium, phosphorous, and other essential plant growth nutrients can rejuvenate depleted soil traits, such as organic carbon, total carbon, total nitrogen, and humic compounds, and improve the physical and chemical properties of soil.

BOMFs and adaptive/resilient crops' combined cultivation can be highly effective in replenishing the agricultural strips downgraded by intense agricultural practices. BOMFs also enhance plant productivity as an outcome of multiple driving factors in long-term applications compared to synthetic fertilizers or traditional compost alone. Furthermore, the application of BOMFs can reduce the amount of chemical fertilizer, consequently reducing the environmental pollution caused by the production, processing, and consumption of chemical fertilizer.

BOMFs can only be used as slow-release fertilizers, and it is difficult to provide sufficient mineral nutrients for the growth of crops in a short period of time. Therefore, it is necessary to use BOMFs in combination with an appropriate amount of chemical fertilizer.

Since BOMFs contain a certain amount of minerals, the mineral types should be mainly basalt, volcanic rock and clay rock, rich in mineral elements, and should be used on the principle of convenience in order to reduce production costs. It must be noted that some rock powders contain rich heavy metals, which are released during mineral weathering and cause environmental pollution. Therefore, the selection of minerals and the amounts in which they are added require sufficient field trials. In short, although the combined application of minerals and organic fertilizers has shown beneficial effects, further research is still required on soil heavy metal pollution, the optimum combination and processing of mineral and organic matter, the mechanism for improving the quality of agricultural products and the evolution of agro-ecological communities.

All living beings in some way are vulnerable to the widespread long-term use of chemicals in agriculture in any form such as fertilisers, pesticides, etc. Agricultural soil has been disrupted by the extensive and disproportionate use of chemicals and putting it back into order will take time and transition.

While we cannot fully prevent the adverse effects of chemical fertilisers at an instant of time, we can definitely reduce the impact by minimising their use and promoting the use of biofertilizers. Biofertilizers will not reduce the use chemical fertilisers but they will improve the soil quality in various ways such as maintaining soil nutrient cycles, soil microbial communities, etc.

Nutrient recovery from organic waste represents a great opportunity to design a new approach in crop fertilization in the framework of the Circular Economy. Nevertheless, recycling nutrients is not enough, as recovered fertilizers should be able to substitute synthetic mineral fertilizers that contain high nutrient concentrations with high nutrient efficiency. A previous paper of ours [181] reported that RF could be effectively obtained thanks to AD and that these RFs were good candidates for replacing SF. In this paper, the LCA approach indicates that producing and using those RFs instead of producing and using SF led to a strong environmental impact reduction.

This result was due above all to the AD process that makes all this possible because of renewable energy production and biological processes modifying the fertilizer properties of digestate. Nevertheless, a correct approach in using RF is mandatory to avoid losing all of the advantages of producing RF because of impacts derived from incorrect RF use. In this way, a well-performed AD process assuring high biological stability of digestate, limiting RF-N_2O emission and RF-NO_3^- leaching, and RF injection limiting NH_3 emissions, as well as using RF at the right time and according to crop requirements should be assured.

Notes from the results that the treatment P2M1B1 gave the highest values, indicating that this level of mineral fertilizer with biological and organic fertilization gave very good results and there is no need for fertilizing with a higher level of mineral fertilizer. In addition, organic fertilization alone or with bio fertilization gave more results than mineral fertilization. Here, it must be said that the added just a mineral fertilizer wasn't enough because the corn requirements for the nutrients are higher than this level, so once the organic and biological fertilizers were added, the qualities improved.

The continuous rise in the global population has translated to a direct increase in the demand for food production. The use of these biofertilizers has been reported to boost the food production rate, and they are a safer farm product for consumers; hence, biofertilizers remains a better alternative for producing safer crops and enhancing global food security.

In recent years, the plant nutrient gap between removal and supply through chemical fertilizer was over 10 million tons. Over-dependence on chemical fertilizers, in terms of both cost and environmental impact, is not a viable strategy in the long run due to the costs involved in setting up fertilizer plants and maintaining production, both in terms of domestic resources and foreign exchange.

Biofertilizers are products that, once adequate information is available to producers and farmers, are likely to be commercially promising in the long run. The use of biofertilizers in the world will not only have an impact on the economic development of sustainable agriculture, but it will also contribute to a sustainable ecosystem and the overall wellbeing of humans.

Based on the survey data of 514 farmers in five provinces and cities around the Bohai Sea in 2019, the influence of organic fertilizer substitution on farmers' technical efficiency was empirically analyzed from the microscopic level. "Policy cognition" and "fertilizer cognition" were instrumental variables in solving the endogenous problem. The heterogeneity analysis was carried out based on the characteristics of farmers. At the same time, the mediating effect of technical efficiency between organic fertilizer substitution of chemical fertilizer and farmer income was tested.

The primary research conclusions are summarized as follows. First, quality and safety awareness, part-time employment level, insurance participation, and training significantly promote the use of organic fertilizers to replace chemical fertilizers. Second, according to the counterfactual analysis, if the farmers who use organic fertilizer substitution do not adopt it, their technical efficiency is 0.571. However, their technical efficiency increases to 0.625 after adoption, a growth rate of 9.46%. Third, the promoting effect of organic fertilizer substitution on technical efficiency has heterogeneity, which is significant among small-scale farmer households with fewer and low-educated family laborers. Fourth, organic fertilizer substitution can indirectly affect farmers' income through the intermediary path of farmers' technical efficiency. Improving technical efficiency is the primary factor for farmers to use organic fertilizer instead of chemical fertilizer to increase income.

Based on the above research conclusions, the policy recommendations of this study are put forward.

First, the promotion of organic fertilizer as alternative fertilizer technology should be encouraged. Based on the positive effect of organic fertilizer substitution of chemical fertilizer on farmers' technical efficiency, the innovation of related technologies should be promoted from multiple perspectives, including organic fertilizer quality and fertilization technology. To increase the promotion of organic fertilizer instead of chemical fertilizer technology, we consider the improvement effect of organic fertilizers on the quality of agricultural products and farmers' income in addition to publicizing the yield increase and environmental protection effects of organic fertilizers. Based on ensuring the long-lasting effect of organic fertilizer, the enthusiasm of farmers to replace chemical fertilizer with organic fertilizer is improved.

Second, efforts should be made to improve the green certification of agricultural products and the agricultural product market supervision system. On the one hand, improving the quality and safety standards of agricultural products is necessary, forcing farmers to change to a green way of production. On the other hand, it is necessary to consolidate and enhance the position of green agricultural products in the market and increase residents' demand for green agricultural products. In addition, making farmers "profitable" for green production provides incentives to use organic fertilizers.

References

1. Khush, G.S. Green revolution: Preparing for the 21st century. *Genome* **1999**, *42*, 646–655.

2. Rohne Till, E. A green revolution in sub-Saharan Africa? The transformation of Ethiopia's agricultural sector. *J. Int. Dev.* **2021**, *33*, 277–315.

3. Martini, E.; Buyer, J.S.; Bryant, D.C.; Hartz, T.K.; Denison, R.F. Yield increases during the organic transition: Improving soil quality or increasing experience? *Field Crop. Res.* **2004**, *86*, 255–266.

4. Dadi, D.; Daba, G.; Beyene, A.; Luis, P.; Van der Bruggen, B. Composting and co-composting of coffee husk and pulp with source-separated municipal solid waste: A breakthrough in valorization of coffee waste. *Int. J. Recycl. Org. Waste Agric.* **2019**, *8*, 263–277.

5. Jones, C.; Jacobsen, J. Nutrient Management Module No.2: Plan Nutrition and Soil Fertility. *Nutr. Manag.* **2001**, *2*, 1–11

6. Willer, H.; Trávnícek, J.; Meier, C.; Schlatter, B. *The World of Organic Agriculture—Statistics & Emerging Trends 2021*; International Federation of Organic Agriculture Movements & Research Institute of Organic Agriculture: Bonn, Germany, 2021.

7. Timsina, J. Can organic sources of nutrients increase crop yields to meet global food demand? *Agronomy* **2018**, *8*, 214.

8. Marinari, S.; Radicetti, E.; Petroselli, V.; Allam, M.; Mancinelli, R. Microbial Indices to Assess Soil Health under Different Tillage and Fertilization in Potato (*Solanum tuberosum* L.) Crop. *Agriculture* **2022**, *12*, 415.

9. Van den Putte, A.; Govers, G.; Diels, J.; Gillijns, K.; Demuzere, M. Assessing the effect of soil tillage on crop growth: A meta-regression analysis on

European crop yields under conservation agriculture. *Eur. J. Agron.* **2010**, *33*, 231–241.

10. Tebrügge, F.; Düring, R.-A.; Du, R.; Tebru, F. Reducing tillage intensity—A review of results from a long-term study in Germany. *Soil Tillage Res.* **1999**, *53*, 15–28.

11. Friedrich, T.; Derpsch, R.; Kassam, A. Overview of the global spread of Conservation Agriculture. *Field Actions Sci. Rep.* **2012**, *6*, 1–7.

12. 8WCCA—8th World Congress on Conservation Agriculture, Bern, Switzerland, 21–23 June 2021. Available online: **https://8wcca.org** (accessed on 14 February 2022).

13. Mancinelli, R.; Marinari, S.; Allam, M.; Radicetti, E. Potential Role of Fertilizer Sources and Soil Tillage Practices to Mitigate Soil CO_2 Emissions in Mediterranean Potato Production Systems. *Sustainability* **2020**, *12*, 8543.

14. Singh, B. Are Nitrogen Fertilizers Deleterious to Soil Health? *Agronomy* **2018**, *8*, 48.

15. Chen, J.; Zhu, R.; Zhang, Q.; Kong, X.; Sun, D. Reduced-tillage management enhances soil properties and crop yields in a alfalfa-corn rotation: Case study of the Songnen Plain, China. *Sci. Rep.* **2019**, *9*, 17064.

16. Amoah-Antwi, C.; Kwiatkowska-Malina, J.; Thornton, S.F.; Fenton, O.; Malina, G.; Szara, E. Restoration of soil quality using biochar and brown coal waste: A review. *Sci. Total Environ.* **2020**, *722*, 137852.

17. Busari, M.A.; Kukal, S.S.; Kaur, A.; Bhatt, R.; Dulazi, A.A. Conservation tillage impacts on soil, crop and the environment. *Int. Soil Water Conserv. Res.* **2015**, *3*, 119–129.

18. Palm, C.; Blanco-Canqui, H.; DeClerck, F.; Gatere, L.; Grace, P. Conservation agriculture and ecosystem services: An overview. *Agric. Ecosyst. Environ.* **2014**, *187*, 87–105.

19. Tscharntke, T.; Grass, I.; Wanger, T.C.; Westphal, C.; Batáry, P. Beyond organic farming—Harnessing biodiversity-friendly landscapes. *Trends Ecol. Evol.* **2021**, *36*, 919–930.

20. Bielińska, E.J.; Mocek-Płóciniak, A. Impact of the tillage system on the soil enzymatic activity. *Arch. Environ. Prot.* **2012**, *38*, 75–82.

21. Rocci, K.S.; Lavallee, J.M.; Stewart, C.E.; Cotrufo, M.F. Soil organic carbon response to global environmental change depends on its distribution between mineral-associated and particulate organic matter: A meta-analysis. *Sci. Total Environ.* **2021**, *793*, 148569. [

22. Dynarski, K.A.; Bossio, D.A.; Scow, K.M. Dynamic Stability of Soil Carbon: Reassessing the "Permanence" of Soil Carbon Sequestration. *Front. Environ. Sci.* **2020**, *8*, 514701.

23. Crystal-Ornelas, R.; Thapa, R.; Tully, K.L. Soil organic carbon is affected by organic amendments, conservation tillage, and cover cropping in organic farming systems: A meta-analysis. *Agric. Ecosyst. Environ.* **2021**, *312*, 107356.

24. Gattinger, A.; Muller, A.; Haeni, M.; Skinner, C.; Fliessbach, A.; Buchmann, N.; Mäder, P.; Stolze, M.; Smith, P.; El-Hage Scialabba, N.; et al. Enhanced top soil carbon stocks under organic farming. *Proc. Natl. Acad. Sci. USA* **2012**, *109*, 18226–18231.

25. Yang, R.; Su, Y.Z.; Wang, T.; Yang, Q. Effect of chemical and organic fertilization on soil carbon and nitrogen accumulation in a newly cultivated farmland. *J. Integr. Agric.* **2016**, *15*, 658–666.

26. Moeskops, B.; Buchan, D.; Van Beneden, S.; Fievez, V.; Sleutel, S.; Gasper, M.S.; D'Hose, T.; De Neve, S. The impact of exogenous organic matter on SOM contents and microbial soil quality. *Pedobiologia* **2012**, *55*, 175–184.

27. Osipitan, O.A.; Radicetti, E. Benefits of sustainable management practices on mitigating greenhouse gas emissions in soybean crop (Glycine max). *Sci. Total Environ.* **2019**, *660*, 1593–1601.

28. Radicetti, E.; Campiglia, E.; Marucci, A.; Mancinelli, R. How winter cover crops and tillage intensities affect nitrogen availability in eggplant. *Nutr. Cycl. Agroecosyst.* **2017**, *108*, 177–194.

29. Mancinelli, R.; Marinari, S.; Brunetti, P.; Radicetti, E.; Campiglia, E. Organic mulching, irrigation and fertilization affect soil CO_2 emission and C storage in tomato crop in the Mediterranean environment. *Soil Tillage Res.* **2015**, *152*, 39–51.

30. Radicetti, E.; Campiglia, E.; Langeroodi, A.S.; Zsembeli, J.; Mendler-Drienyovszki, N.; Mancinelli, R. Soil carbon dioxide emissions in eggplants based on cover crop residue management. *Nutr. Cycl. Agroecosyst.* **2020**, *118*, 39–55.

31. Papp, R.; Marinari, S.; Moscatelli, M.C.; van der Heijden, M.G.A.; Wittwer, R.; Campiglia, E.; Radicetti, E.; Mancinelli, R.; Fradgley, N.; Pearce, B.; et al. Short-term changes in soil biochemical properties as affected by subsidiary crop cultivation in four European pedo-climatic zones. *Soil Tillage Res.* **2018**, *180*, 126–136.

32. Hammed, T.B.; Oloruntoba, E.O.; Ana, G.R.E.E. Enhancing growth and yield of crops with nutrient-enriched organic fertilizer at wet and dry seasons in ensuring climate-smart agriculture. *Int. J. Recycl. Org. Waste Agric.* **2019**, *8*, 81–92.

33. Szostek, M.; Szpunar-Krok, E.; Pawlak, R.; Stanek-Tarkowska, J.; Ilek, A. Effect of Different Tillage Systems on Soil Organic Carbon and Enzymatic Activity. *Agronomy* **2022**, *12*, 208.

34. Farooq, M.; Flower, K.C.; Jabran, K.; Wahid, A.; Siddique, K.H.M. Crop yield and weed management in rainfed conservation agriculture. *Soil Tillage Res.* **2011**, *117*, 172–183.

35. Knapp, S.; van der Heijden, M.G.A. A global meta-analysis of yield stability in organic and conservation agriculture. *Nat. Commun.* **2018**, *9*, 3632.

36. Schmidt, R.; Gravuer, K.; Bossange, A.V.; Mitchell, J.; Scow, K. Long-term use of cover crops and no-till shift soil microbial community life strategies in agricultural soil. *PLoS ONE* **2018**, *13*, e0192953.

37. Wang, X.; Yan, J.; Zhang, X.; Zhang, S.; Chen, Y. Organic manure input improves soil water and nutrients use for sustainable maize (*Zea mays* L.) productivity on the Loess Plateau. *PLoS ONE* **2020**, *15*, e0238042.

38. Nouraein, M.; Skataric, G.; Spalevic, V.; Dudic, B.; Gregus, M. Short-term effects of tillage intensity and fertilization on sunflower yield, achene quality, and soil physicochemical properties under semi-arid conditions. *Appl. Sci.* **2019**, *9*, 5482.

39. Ordoñez-Morales, K.D.; Cadena-Zapata, M.; Zermeño-González, A.; Campos-Magaña, S. Effect of tillage systems on physical properties of a clay loam soil under oats. *Agriculture* **2019**, *9*, 62.

40. Ramakrishna Parama, V.R.; Munawery, A. Sustainable soil nutrient management. *J. Indian Inst. Sci.* **2012**, *92*, 1–16.

41. Liverpool-Tasie, L.S.O.; Omonona, B.T.; Sanou, A.; Ogunleye, W.O. Is increasing inorganic fertilizer use for maize production in SSA a profitable proposition? Evidence from Nigeria. *Food Policy* **2017**, *67*, 41–51.

42. Du, Y.; Cui, B.; Zhang, Q.; Wang, Z.; Sun, J.; Niu, W. Effects of manure fertilizer on crop yield and soil properties in China: A meta-analysis. *Catena* **2020**, *193*, 104617.

43. Hijbeek, R.; van Ittersum, M.K.; ten Berge, H.F.M.; Gort, G.; Spiegel, H.; Whitmore, A.P. Do organic inputs matter—A meta-analysis of additional yield effects for arable crops in Europe. *Plant Soil* **2017**, *411*, 293–303.

44. Mancinelli, R.; Muleo, R.; Marinari, S.; Radicetti, E. How soil ecological intensification by means of cover crops affects nitrogen use efficiency in pepper cultivation. *Agriculture* **2019**, *9*, 145.

45. De Ponti, T.; Rijk, B.; Van Ittersum, M.K. The crop yield gap between organic and conventional agriculture. *Agric. Syst.* **2012**, *108*, 1–9.

46. Sun, T.; Wang, Y.; Hui, D.; Jing, X.; Feng, W. Soil properties rather than climate and ecosystem type control the vertical variations of soil organic carbon, microbial carbon, and microbial quotient. *Soil Biol. Biochem.* **2020**, *148*, 107905.

47. Chivenge, P.; Vanlauwe, B.; Six, J. Does the combined application of organic and mineral nutrient sources influence maize productivity? A meta-analysis. *Plant Soil* **2011**, *342*, 1–30.

48. Lin, Y.; Watts, D.B.; Van Santen, E.; Cao, G. Influence of poultry litter on crop productivity under different field conditions: A meta-analysis. *Agron. J.* **2018**, *110*, 807–818.

49. Allam, M.; Radicetti, E.; Petroselli, V.; Mancinelli, R. Meta-Analysis Approach to Assess the Effects of Soil Tillage and Fertilization Source under Different Cropping Systems. *Agriculture* **2021**, *11*, 823.

50. Liu, Z.; Cao, S.; Sun, Z.; Wang, H.; Qu, S.; Lei, N.; He, J.; Dong, Q. Tillage effects on soil properties and crop yield after land reclamation. *Sci. Rep.* **2021**, *11*, 4611.

51. Obour, A.K.; Holman, J.D.; Simon, L.M.; Schlegel, A.J. Strategic tillage effects on crop yields, soil properties, and weeds in dryland no-tillage systems. *Agronomy* **2021**, *11*, 662.

52. Rusinamhodzi, L.; Corbeels, M.; Van Wijk, M.T.; Rufino, M.C.; Nyamangara, J.; Giller, K.E. A meta-analysis of long-term effects of conservation agriculture on maize grain yield under rain-fed conditions. *Agron. Sustain. Dev.* **2011**, *31*, 657–673.

53. Wei, W.; Yan, Y.; Cao, J.; Christie, P.; Zhang, F.; Fan, M. Effects of combined application of organic amendments and fertilizers on crop yield and soil organic matter: An integrated analysis of long-term experiments. *Agric. Ecosyst. Environ.* **2016**, *225*, 86–92.

54. Campiglia, E.; Mancinelli, R.; De Stefanis, E.; Pucciarmati, S.; Radicetti, E. The long-term effects of conventional and organic cropping systems, tillage managements and weather conditions on yield and grain quality of durum wheat (*Triticum durum* Desf.) in the Mediterranean environment of Central Italy. *Field Crop. Res.* **2015**, *176*, 34–44.

55. Zingore, S.; Delve, R.J.; Nyamangara, J.; Giller, K.E. Multiple benefits of manure: The key to maintenance of soil fertility and restoration of depleted sandy soils on African smallholder farms. *Nutr. Cycl. Agroecosystems* **2008**, *80*, 267–282.

56. Rurangwa, E.; Vanlauwe, B.; Giller, K.E. Benefits of inoculation, P fertilizer and manure on yields of common bean and soybean also increase yield of subsequent maize. *Agric. Ecosyst. Environ.* **2018**, *261*, 219–229.

57. Faligowska, A.; Kalembasa, S.; Kalembasa, D.; Panasiewicz, K.; Szymanska, G.; Ratajczak, K.; Skrzypczak, G. The Nitrogen fixation and yielding of pea in different soil tillage Systems. *Agronomy* **2022**, *12*, 352.

58. Cooper, J.M.; Baranski, M.; Nobel de Lange, M.; Barberi, P.; Fliessbach, A.; Peigne, J.; Berner, A.; Brock, C.; Casagrande, M.; Crowley, O.; et al. Effects of reduced tillage in organic farming on yield, weeds and soil carbon: Meta-analysis results from the TILMAN-ORG project. In Proceedings of the 4th ISOFAR Scientific Conference. 'Building Organic Bridges', at the Organic World Congress 2014, Istanbul, Turkey, 13–15 October 2014; Volume 4.

59. Idowu, O.J.; Sultana, S.; Darapuneni, M.; Beck, L.; Steiner, R.; Omer, M. Tillage effects on cotton performance and soil quality in an irrigated arid cropping system. *Agriculture* **2020**, *10*, 531.

60. Orzech, K.; Wanic, M.; Załuski, D. The effects of soil compaction and different tillage systems on the bulk density and moisture content of soil and the yields of winter oilseed rape and cereals. *Agriculture* **2021**, *11*, 666.

61. Liu, Q.; Xu, H.; Mu, X.; Zhao, G.; Gao, P.; Sun, W. Effects of different fertilization regimes on crop yield and soil water use efficiency of millet and soybean. *Sustainability* **2020**, *12*, 4125.

62. Celik, I.; Ortas, I.; Kilic, S. Effects of compost, mycorrhiza, manure and fertilizer on some physical properties of a Chromoxerert soil. *Soil Tillage Res.* **2004**, *78*, 59–67.

63. Nyamangara, J.; Gotosa, J.; Mpofu, S.E. Cattle manure effects on structural stability and water retention capacity of a granitic sandy soil in Zimbabwe. *Soil Tillage Res.* **2001**, *62*, 157–162.

64. Liu, C.A.; Li, F.R.; Zhou, L.M.; Zhang, R.H.; Jia, Y.; Lin, S.L.; Wang, L.J.; Siddique, K.H.M.; Li, F.M. Effect of organic manure and fertilizer on soil

water and crop yields in newly-built terraces with loess soils in a semi-arid environment. *Agric. Water Manag.* **2013**, *117*, 123–132.

65. Osman, K.T. Soil resources and soil degradation. In *Soil Degradation, Conservation and Remediation*; Springer: Berlin/Heidelberg, Germany, 2014; pp. 1–43. [**Google Scholar**]

66. Lin, W.W.; Lin, M.; Zhou, H.; Wu, H.; Li, Z.W.; Lin, W.X. The effects of chemical and organic fertilizer usage on rhizosphere soil in tea orchards. *PLoS ONE* **2019**, *14*, e0217018. [**Google Scholar**] [**CrossRef**] [**PubMed**]

67. Xiao, L.; Sun, Q.; Yuan, H.; Lian, B.A. Practical soil management to improve soil quality by applying mineral organic fertilizer. *Acta Geochim.* **2017**, *36*, 198–204. [**Google Scholar**] [**CrossRef**]

68. Xiao, L.; Sun, Q.; Yuan, H.; Li, X.; Chu, Y.; Ruan, Y.; Lian, B. A feasible way to increase carbon sequestration by adding dolomite and K-feldspar to soil. *Cogent Geosci.* **2016**, *2*, 1205324. [**Google Scholar**] [**CrossRef**]

69. Sun, Q.; Ruan, Y.; Chen, P.; Wang, S.; Liu, X.; Lian, B. Effects of mineral-organic fertilizer on the biomass of green Chinese cabbage and potential carbon sequestration ability in karst areas of Southwest China. *Acta Geochim.* **2019**, *38*, 430–439. [**Google Scholar**] [**CrossRef**]

70. Slessarev, E.W.; Chadwick, O.A.; Sokol, N.W.; Nuccio, E.E.; Pett-Ridge, J. Rock weathering controls the potential for soil carbon storage at a continental scale. *Biogeochemistry.* **2021**. [**Google Scholar**] [**CrossRef**]

71. Wang, Q.; Wang, R.; He, L.; Sheng, X. Location-related differences in weathering behaviors and populations of culturable rock-weathering bacteria along a hillside of a rock mountain. *Microb. Ecol.* **2017**, *73*, 838–849. [**Google Scholar**] [**CrossRef**]

72. Yang, X.; Lian, B.; Zhu, X.L.; An, Y.L.; Chen, J.; Zhu, L.J. Effects of adding potassium-bearing mineral powder on nitrogen, potassium and potassium contents of chicken manure compost. *Earth Environ.* **2012**, *40*, 286–292. [**Google Scholar**]

73. Ditta, A.; Muhammad, J.; Imtiaz, M.; Mehmood, S.; Qian, Z.; Tu, S. Application of rock phosphate enriched composts increases nodulation, growth and yield of chickpea. *Int. J. Recycl. Org. Waste Agric.* **2018**, *7*, 33–40. [**Google Scholar**] [**CrossRef**][**Green Version**]

74. Niamat, B.; Naveed, M.; Ahmad, Z.; Yaseen, M.; Ditta, A.; Mustafa, A.; Xu, M. Calcium-enriched animal manure alleviates the adverse effects of salt stress on growth, physiology and nutrients homeostasis of *Zea mays* L. *Plants* **2019**, *8*, 480. [**Google Scholar**] [**CrossRef**][**Green Version**]

75. Masruroh, A.; Minardi, S. Rock phosphate, zeolite and quail manure to enhance potassium uptake and yield of soybean on alfisols. *Asian J. Soil Sci. Plant Nutr.* **2019**, *5*, 1–9. [**Google Scholar**] [**CrossRef**][**Green Version**]

76. Li, Y.; Liu, X.M.; Zhang, L.; Xie, Y.H.; Cai, X.L.; Wang, S.J.; Lian, B. Effects of short-term application of chemical and organic fertilizers on bacterial diversity of cornfield soil in a karst area. *J. Soil Sci. Plant Nutr.* **2020**, *20*, 2048–2058. [**Google Scholar**] [**CrossRef**]

77. Thorley, R.M.S.; Taylor, L.L.; Banwart, S.A.; Leake, J.R.; Beerling, D.J. The role of forest trees and their mycorrhizal fungi in carbonate rock wearthering and its significance for global carbon cycling. *Plant Cell Environ.* **2015**, *38*, 1947–1961. [**Google Scholar**] [**CrossRef**][**Green Version**]

78. Zheng, X.; Zhu, Y.; Wang, Z.; Zhang, H.; Chen, M.; Chen, Y.; Liu, B. Effects of a novel bio-organic fertilizer on the composition of rhizobacterial communities and bacterial wilt outbreak in a continuously mono-cropped

tomato field. *Appl. Soil Ecol.* **2020**, *156*, 103717. [**Google Scholar**] [**CrossRef**]

79. Han, Y.; Feng, G.; Swaney, D.P.; Dentener, F.; Koeble, R.; Ouyang, Y.; Gao, W. Global and regional estimation of net anthropogenic nitrogen inputs (NANI). *Geoderma* **2020**, *361*, 114066. [**Google Scholar**] [**CrossRef**]

80. Food and Agriculture Organization. Available online: **http://fenix.fao.org/faostat/internal/en/#data** (accessed on 18 October 2021).

81. Chianu, J.N.; Mairura, F. Mineral fertilizers in the farming systems of sub-Saharan Africa—A review. *Agron. Sustain. Dev.* **2012**, *32*, 545–566. [**Google Scholar**] [**CrossRef**][**Green Version**]

82. Geisseler, D.; Scow, K.M. Long-term effects of mineral fertilizers on soil microorganisms—A review. *Soil Biol. Biochem.* **2014**, *75*, 54–63. [**Google Scholar**] [**CrossRef**]

83. Lambrecht, I.; Vanlauwe, B.; Merckx, R.; Maertens, M. Understanding the process of agricultural technology adoption: Mineral fertilizer in eastern DR Congo. *World Dev.* **2014**, *59*, 132–146. [**Google Scholar**] [**CrossRef**]

84. Egodawatta, W.C.P.; Sangakkara, U.R.; Stamp, P. Impact of green manure and mineral fertilizer inputs on soil organic matter and crop productivity in a sloping landscape of Sri Lanka. *Field Crop. Res.* **2012**, *129*, 21–27. [**Google Scholar**] [**CrossRef**]

85. Du, N.X.T. Effects of green manures during fallow on moisture and nutrients of soil and winter wheat yield on the Loss Plateau of China. *Emir. J. Food Agric.* **2017**, *29*, 978–987. [**Google Scholar**]

86. Flores-Félix, J.D.; Menéndez, E.; Rivas, R.; de la Encarnación Velázquez, M. Future perspective in organic farming fertilization: Management and product.

In *Organic Farming*; Woodhead Publishing: Sawston, UK, 2018; pp. 269–315. [**Google Scholar**]

87. Zhu, B.; Wang, T.; You, X.; Gao, M.R. Nutrient release from weathering of purplish rocks in the Sichuan Basin, China. *Pedosphere* **2008**, *18*, 257–264. [**Google Scholar**] [**CrossRef**]

88. Bieluczyk, W.; Piccolo, M.; Pereira, M.G. Integrated farming systems influence soil organic matter dynamics in southeastern Brazil. *Geoderma* **2020**, *371*, 114368. [**Google Scholar**] [**CrossRef**]

89. Bronick, C.J.; Lal, R. Soil structure and management: A review. *Geoderma* **2005**, *124*, 3–22. [**Google Scholar**] [**CrossRef**]

90. Lian, B.; Chen, Y.; Zhu, L.; Yang, R. Effect of microbial weathering on carbonate rocks. *Earth Sci. Front.* **2008**, *15*, 90–99. [**Google Scholar**] [**CrossRef**]

91. Yang, Y.; Syed, S.; Mao, S.; Li, Q.; Ge, F.; Lian, B.; Lu, C.M. Bio-organic–mineral fertilizer can remediate chemical fertilizer-oversupplied soil: Purslane planting as an example. *J. Soil Sci. Plant Nutr.* **2020**, *20*, 892–900. [**Google Scholar**] [**CrossRef**]

92. Li, M.; Li, Q.; Yun, J.; Yang, X.; Wang, X.; Lian, B.; Lu, C.M. Bio-organic-mineral fertilizer can improve soil quality and promote the growth and quality of water spinach. *Can. J. Soil Sci.* **2017**, *97*, 552–560. [**Google Scholar**] [**CrossRef**][**Green Version**]

93. Li, R.Y.; Pang, Z.Q.; Zhou, Y.M.; Fyumah, N.; Hu, C.H.; Lin, W.X.; Yuan, Z.N. Metagenomic analysis exploring taxonomic and functional diversity of soil microbial communities in sugarcane fields applied with organic fertilizer. *BioMed Res. Int.* **2020**, *2020*, 9381506. [**Google Scholar**] [**CrossRef**]

94. Sindhu, S.S.; Parmar, P.; Phour, M.; Sehrawat, A. Potassium-solubilizing microorganisms (KSMs) and its effect on plant growth improvement. In *Potassium Solubilizing Microorganisms for Sustainable Agriculture*; Springer: Berlin/Heidelberg, Germany, 2016; pp. 171–185. [**Google Scholar**]

95. Etesami, H.; Emami, S.; Alikhani, H.A. Potassium solubilizing bacteria (KSB): Mechanisms, promotion of plant growth, and future prospects-A review. *J. Soil Sci. Plant Nutr.* **2017**, *17*, 897–911. [**Google Scholar**] [**CrossRef**]

96. Ribeiro, I.D.A.; Volpiano, C.G.; Vargas, L.K.; Granada, C.E.; Lisboa, B.B.; Passaglia, L.M.P. Use of mineral weathering bacteria to enhance nutrient availability in crops: A review. *Front. Plant Sci.* **2020**, *11*, 590774. [**Google Scholar**] [**CrossRef**]

97. Berde, C.V.; Gawde, S.S.; Berde, V.B. Potassium solubilization: Mechanism and functional impact on plant growth. In *Soil Microbiomes for Sustainable Agriculture*; Springer: Berlin/Heidelberg, Germany, 2021. [**Google Scholar**]

98. Goswami, D.; Parmar, S.; Vaghela, H.; Dhandhukia, P.; Thakker, J.N. Describing *Paenibacillus mucilaginosus* strain N3 as an efficient plant growth promoting rhizobacteria (PGPR). *Cogent Food Agric.* **2015**, *1*, 1000714. [**Google Scholar**] [**CrossRef**]

99. Wang, P.; Wu, S.H.; Wen, M.X.; Wang, Y.; Wu, Q.S. Effects of combined inoculation with *rhizophagus intraradices* and *Paenibacillus mucilaginosus* on plant growth, root morphology, and physiological status of trifoliate orange (*Poncirus trifoliata* L. Raf.) seedlings under different levels of phosphorus. *Sci. Hortic.* **2016**, *205*, 97–105. [**Google Scholar**] [**CrossRef**]

100. Chen, Y.H.; Yang, X.Z.; Li, Z.; An, X.H.; Ma, R.P.; Li, Y.Q.; Cheng, C.G. Efficiency of potassium-solubilizing *paenibacillus mucilaginosus* for the

growth of apple seedling. *J. Integr. Agric.* **2020**, *19*, 2458–2469. [**Google Scholar**] [**CrossRef**]

101. Basak, B.B. Recycling of waste biomass and mineral powder for preparation of potassium-enriched compost. *J. Mater. Cycles Waste Manag.* **2018**, *20*, 1409–1415. [**Google Scholar**] [**CrossRef**]

102. Basak, B.B.; Maity, A.; Biswas, D.R. *Cycling of natural sources of phosphorus and potassium for environmental sustainability. Biogeochemical Cycles: Ecological Drivers and Environmental Impact*; Advancing Earth and Space Science: Washington, DC, USA, 2020; pp. 285–299. [**Google Scholar**]

103. Yu, X.D.; Wang, B.; Lian, B. Effect of organic fermentative fertilizer made from potassium-bearing rocks in the growth of *amaranth mangostanus*. *Soil Fertil. Sci. China.* **2011**, *2*, 61–64. [**Google Scholar**]

104. Chen, P.; Ruan, Y.L.; Wang, S.J.; Liu, X.M.; Lian, B. Effects of organic mineral fertiliser on heavy metal migration and potential carbon sink in soils in a karst region. *Acta Geochim.* **2017**, *36*, 539–543. [**Google Scholar**] [**CrossRef**]

105. Belov, S.V.; Danyleiko, Y.K.; Glinushkin, A.P.; Kalinitchenko, V.P.; Egorov, A.V.; Sidorov, V.A.; Izmailov, A.Y. An activated potassium phosphate fertilizer solution for stimulating the growth of agricultural plants. *Front. Phys.* **2021**, *8*, 618320. [**Google Scholar**] [**CrossRef**]

106. Zoca, S.M.; Penn, C. An important tool with no instruction manual: A review of gypsum use in agriculture. *Adv. Agron.* **2017**, *144*, 1–44. [**Google Scholar**]

107. Noordin, W.D.; Zulkefly, S.; Shamshuddin, J.; Hanafi, M.M. Improving soil chemical properties and growth performance of *Hevea brasiliensis* through basalt application. *Int. Proc. IRC* **2017**, *2017*, 308–323. [**Google Scholar**] [**CrossRef**]

108. Sheldrick, W.; Syers, J.K.; Lingard, J. Contribution of livestock excreta to nutrient balances. *Nutr. Cycl. Agroecosyst.* **2003**, *66*, 119–131. [**Google Scholar**] [**CrossRef**]

109. Manyuchi, M.M.; Phiri, A.; Muredzi, P.; Chitambwe, T. Comparison of vermicompost and vermiwash bio-fertilizers from vermicomposting waste corn pulp. *Int. J. Biol. Biomol. Agric. Food Biotechnol. Eng.* **2013**, *7*, 389–392. [**Google Scholar**]

110. Liu, H.T.; Chen, T.B.; Zhen, T.D.; Gao, D.; Lei, M. Comparative analysis of energy consumption, input cost and environmental benefit of organic fertilizer and chemical fertilizer production—take sludge composting to produce organic fertilizer as an example. *Ecol. Environ. Sci.* **2010**, *19*, 1000–1003. [**Google Scholar**]

111. Nacke, H.; Junior, G.; Schwantes, J.R.; Nava, L. Productivity and yield components of corn fertilized with different sources and levels of zinc. *Span. J. Rural Dev.* **2011**, *2*, 71–79. [**Google Scholar**] [**CrossRef**]

112. Vezzani, F.M.; Anderson, C.; Meenken, E.; Gillespie, R.; Peterson, M.; Beare, M.H. The importance of plants to development and maintenance of soil structure, microbial communities and ecosystem functions. *Soil Tillage Res.* **2018**, *175*, 139–149. [**Google Scholar**] [**CrossRef**]

113. Lazcano, C.; Gómez-Brandón, M.; Revilla, P.; Domínguez, J. Short-term effects of organic and inorganic fertilizers on soil microbial community structure and function. *Biol. Fertil. Soils* **2013**, *49*, 723–733. [**Google Scholar**] [**CrossRef**]

114. Huang, P.; Zhang, J.B.; Xin, X.L.; Zhu, A.N.; Zhang, C.Z.; Zhu, Q.G.; Wu, S.J. Proton accumulation accelerated by heavy chemical nitrogen fertilization and its long-term impact on acidifying rate in a typical arable soil in the

Huang-Huai-Hai Plain. *J. Integr. Agric.* **2015**, *14*, 148–157. [**Google Scholar**] [**CrossRef**]

115. Zamanian, K.; Zarebanadkouki, M.; Kuzyakov, Y. Nitrogen fertilization raises CO_2 efflux from inorganic carbon: A global assessment. *Glob. Chang. Biol.* **2018**, *24*, 2810–2817. [**Google Scholar**] [**CrossRef**]

116. Raza, S.; Miao, N.; Wang, P.; Ju, X.; Chen, Z.; Zhou, J.; Kuzyakov, Y. Dramatic loss of inorganic carbon by nitrogen-induced soil acidification in Chinese croplands. *Glob. Chang. Biol.* **2020**, *26*, 3738–3751. [**Google Scholar**] [**CrossRef**]

117. Datta, S.P.; Rattan, R.K.; Chandra, S. Labile soil organic carbon, soil fertility, and crop productivity as influenced by manure and mineral fertilizers in the tropics. *J. Plant Nutr. Soil Sci.* **2010**, *173*, 715–726. [**Google Scholar**] [**CrossRef**]

118. Shen, Z.; Ruan, Y.; Chao, X.; Zhang, J.; Li, R.; Shen, Q. Rhizosphere microbial community manipulated by 2 years of consecutive biofertilizer application associated with banana fusarium wilt disease suppression. *Biol. Fertil. Soils* **2015**, *51*, 553–562. [**Google Scholar**] [**CrossRef**]

119. Verma, R.; Maurya, B.R.; Meena, V.S.; Dotaniya, M.L.; Deewan, P. Microbial dynamics as influenced by bio-organics and mineral fertilizer in alluvium soil of Varanasi, India. *Int. J. Curr. Microbiol. Appl. Sci.* **2017**, *6*, 1516–1524. [**Google Scholar**] [**CrossRef**][**Green Version**]

120. Verma, R.; Maurya, B.R.; Meena, V.S.; Dotaniya, M.L.; Deewan, P.; Jajoria, M. Enhancing production potential of cabbage and improves soil fertility status of Indo-Gangetic Plain through application of bio-organics and mineral fertilizer. *Int. J. Curr. Microbiol. Appl. Sci.* **2017**, *6*, 301–309. [**Google Scholar**]

121. Mahajan, S.; Kanwar, S.S.; Sharma, S.P. Long-term effect of mineral fertilizers and amendments on microbial dynamics in an alfisol of Western Himalayas. *Indian J. Microbiol.* **2007**, *47*, 86–89. [**Google Scholar**] [**CrossRef**][**Green Version**]

122. Diacono, M.; Montemurro, F. Long-term effects of organic amendments on soil fertility—A review. *Agron. Sustain. Dev.* **2010**, *30*, 401–422. [**Google Scholar**] [**CrossRef**][**Green Version**]

123. Huang, N.; Wang, W.W.; Yao, Y.L.; Zhu, F.X.; Wang, W.P.; Chang, X.J. The influence of different concentrations of bio-organic fertilizer on cucumber fusarium wilt and soil microflora alterations. *PLoS ONE* **2017**, *12*, e0171490. [**Google Scholar**] [**CrossRef**]

124. Yang, W.H.; Chang, J.; Wang, S.S.; Zhou, B.Q.; Mao, Y.L.; Rensing, C.; Xing, S.H. Influence of biochar and biochar-based fertilizer on yield, quality of tea and microbial community in an acid tea orchard soil. *Appl. Soil Ecol.* **2021**, *166*, 104005. [**Google Scholar**] [**CrossRef**]

125. Chen, J. 2007. Rapid urbanization in China: a real challenge to soil protection and food security. Catena 69:1-15

126. Ashraf, M.A., Maah, M.J., Yusoff, I. 2014. Soil contamination, risk assessment and remediation. In: Hernandez-Soriano MC (ed) Environmental risk assessment of soil contamination. INTECH, Rijeka, 3-56

127. Masindi, V., Muedi, K. 2018. Environmental contamination by heavy metals, heavy metals. Hosam El-Din M. Saleh & Refaat F. Aglan, IntechOpen

128. Edwards, C.A. 2002. Assessing the effects of environmental pollutants on soil organisms, communities, processes

129. Liu, H., Probst, A., Liao, B. 2005. Metal contamination of soils and crops affected by the Chenzhou lead/zinc mine spill (Hunan, China). Sci Total Environ 339:153-166

130. Mclaughlin, M.J., Tiller, K.G., Naidu, R., Stevens, D.P. 1996. Review: the behaviour and environmental impact of contaminants in fertilizers. Aust J Soil Res 34:1-54

131. Calace, N., Fiorentini, F., Petronio, B.M., Pietroletti, M. 2001. Effects of acid rain on soil humic compounds. Talanta. 54:837-846

132. Zhang, J.E., Ouyang, Y., Ling. D.J. 2007. Impacts of simulated acid rain on cation leaching from the Latosol in south China. Chemosphere. 67:2131-2137

133. Neina, D. 2019. The Role of Soil pH in Plant Nutrition and Soil Remediation. Appl. Environ. Soil Sci. 5794869

134. Wang, Q.R., Cui, Y.S., Liu, X.M., Dong, Y.T., Christie, P. 2003. Soil contamination and plant uptake of heavy metals at polluted sites in China. J Environ Sci Health 38:823-838

135. Zhao, Y.F., Shi, X.Z., Huang, B., Yu, D.S., Wang, H.J., Sun, W.X., Öboern, I., Blombäck. K. 2007. Spatial distribution of heavy metals in agricultural soils of an industry-based peri-urban area in Wuxi, China. Pedosphere.17:44-51

136. Giller, K.E., Witter, E., Mcgrath, S.P. 1998. Toxicity of heavy metals to microorganisms and microbial processes in agricultural soils: a review. Soil Biol Biochem. 30:1389-1414

137. Lenart-Boroń, A., and Boroń, P. 2014. The effect of industrial heavy metal pollution on microbial abundance and diversity in soils–a review. Actinomycetes, 1012, 107-108

138. Bradl, H.B. 2004. Adsorption of heavy metal ions on soils and soils constituents. J Colloid Interface Sci. 277(1), 1-18

139. Mclaughlin, M.J., Tiller, K.G., Naidu, R., Stevens, D.P. 1996. Review: the behaviour and environmental impact of contaminants in fertilizers. Aust J Soil Res 34:1-54

140. Arias-Estévez, M., López-Periago, E., Martínez-Carballo, E., Simal-Gándara, J., Mejuto, J.C., García-Río, L. 2008. The mobility and degradation of pesticides in soils and the pollution of groundwater resources. Agric Ecosyst Environ 123:247-260

141. Doran, J.W. 2002. Soil health and global sustainability: translating science into practice. Agric Ecosyst Environ 88:119-127

142. Tahat, M., Alananbeh, K., Othman, Y., Leskovar, D. 2020. Soil Health and Sustainable Agriculture. Sustainability. 12, 4859

143. Rojas, R.V., Achouri, M., Maroulis, J. Caon, L. 2016. Healthy soils: a prerequisite for sustainable food security. Environ Earth Sci 75, 180

144. Black, R. 2003. Micronutrient deficiency—an underlying cause of morbidity and mortality. No. 81 (2) World Health Organization

145. Shetty, P. 2009. Incorporating nutritional considerations when addressing food insecurity. Food Security 1:431-440

146. Charles, H., Godfray, J., Beddington, J.R., Crute, I.R., Haddad, L. 2010. Food security: the challenge of feeding 9 billion people. Science 327:812-817

147. FAO, WHO (2014) Rome declaration on nutrition. second international conference on nutrition. Rome, Italy

148. Clair, S.B., Lynch, J.P. 2010. The opening of pandora's box: climate change impacts on soil fertility and crop nutrition in developing countries. Plant Soil 335:101-115

149. Li, D. P., and Wu, Z. J. 2008. Impact of chemical fertilizers application on soil ecological environment. Chinese Journal of Applied Ecology, 19, 1158-1165

150. Weisskopf, P., Reiser, R., Rek, J., Oberholzer, H.R. 2010. Effect on different compaction impacts and varying subsequent management practices on soil structure, air regime and microbiological parameters. Soil Till. Research111,65-74

151. Mari, G. R., Ji. Changying, Jun Zhou. 2008. Effects of soil compaction on soil physical properties and nitrogen, phosphorus, potassium uptake in wheat plants. J. Transactions of the CSAE, 24(1): 74-79

152. Batey, T. 2009. Soil compaction and soil management -a review. Soil Use and Management. 25(4): 335-345

153. Blanco, C. H., C. H. Gantzer, S. H. Anderson, E. E. Alberts, and F. Ghidey. 2002. Saturated hydraulic conductivity and its impact on simulated runoff for claypan soils. Soil Sci. Soc. Am. J. 66, 1596-1602

154. Rannik, K. 2009. Soil compaction effects on soil bulk density and penetration resistance and growth of spring barley (Hordeum vulgare L.). J. Acta Agriculturae Scandinavica. Plant Soil Science. 59 (3): 265-272

155. Barzegar, A. R., Nadian, H., Heidari, F., Herbert, S. J., Hashemi, A. M. 2006. Interaction of soil compaction, phosphorus and zinc on clover growth and accumulation of phosphorus. Soil & Tillage Res. 87: 155-162

156. Beylich, A., Oberholzer, H. R., Schrader, S., Hoper, H., Wilke. B. M. 2010. Evaluation of soil compaction effects on soil biota and soil biological processes in soils. Soil & Tillage Res. 109(2): 133-143

157. Dexter, A. R., Richard, G., Arrouays, D., Czyz, E. A., Jolivet, C., Duval, O. 2008. Complexed organic matter controls soil physical properties. Geoderma. 144, 620-627

158. Celik, I., Gunal, H., Budak, M., Akpinar, C. 2010. Effects of long-term organic and mineral fertilizers on bulk density and penetration resistance in semiarid Mediterranean soil conditions. Geoderma. 160(2): 236-243

159. Chandini, K. R., Kumar, R., Prakash, O. 2019. The impact of chemical fertilizers on our environment and ecosystem. Research trends in environmental Sciences, 2nd edition, pp 69-86

160. Sonmez Kaplan, M., Sonmez, S. 2007. An investigation of seasonal changes in nitrate contents of soils and irrigation waters in greenhouses located in Antalya-Demre region. Asian Journal of Chemistry. 19(7):5639

161. Savci, S. 2012. Investigation of effect of chemical fertilizers on environment. Apcbee Procedia.1, 287-292

162. Bisht, N., Tiwari, S., Singh, P.C., Niranjan, A., Chauhan, P.S., 2019. A multifaceted rhizobacterium Paenibacillus lentimorbus alleviates nutrient deficiency-induced stress in Cicer arietinum L. Microbiol. Res. 223-225, 110-119

163. Bisht, N. and Chauhan, P. S. 2020. Comparing the growth-promoting potential of Paenibacillus lentimorbus and Bacillus amyloliquefaciens in Oryza sativa L. var. Sarju-52 under suboptimal nutrient conditions. Plant Physiology and Biochemistry, 146, 187-197

164. Jacoby, R., Peukert, M., Succurro, A., Koprivova, A.,Kopriva, S. 2017. The role of soil microorganisms in plant mineral nutrition-current knowledge and future directions. Front. Plant Sci. 8:1617

165. Bargaz, A., Lyamlouli, K., Chtouki, M., Zeroual, Y., Dhiba, D. 2018. Soil microbial resources for improving fertilizers efficiency in an integrated plant nutrient management system. Front. Micribiol. 9:1606

166. Liang, R., Hou, R., Li, J., Lyu, Y., Hang, S., Gong, H., Ouyang, Z. 2020. Effects of Different Fertilizers on Rhizosphere Bacterial Communities of Winter Wheat in the North China Plain. Agronomy.10, 93

167. Wu, L., Jiang, Y., Zhao, F. He, X., Liu, H., Yu, K. 2020. Increased organic fertilizer application and reduced chemical fertilizer application affect the soil properties and bacterial communities of grape rhizosphere soil. Sci Rep 10, 9568

168. Bisht N., Chauhan P.S. 2020. Microorganisms in Maintaining Food and Energy Security in a World of Shifting Climatic Conditions. In: Singh P., Singh R., Srivastava V. (eds) Contemporary Environmental Issues and Challenges in Era of Climate Change. Springer, Singapore

169. Ramasamy,M., Geetha, T., Yuvaraj, M. 2020. Role of Biofertilizers in Plant Growth and Soil Health, Nitrogen Fixation, Everlon Cid Rigobelo and Ademar Pereira Serra, IntechOpen, DOI: 10.5772/intechopen.87429

170. Bhardwaj, D., Ansari, M. W., Sahoo, R. K., Tuteja, N. 2014. Biofertilizers function as key player in sustainable agriculture by improving soil fertility, plant tolerance and crop productivity. Microb. Cell Fact. 13:66. doi: 10.1186/1475-2859-13-66

171. Ahemad, M., Kibret, M. 2014. Mechanisms and applications of plant growth promoting rhizobacteria: current perspective. JKSUS 26, 1-20

172. Rockström, J.; Gaffney, O.; Thunberg, G. Breaking Boundaries: The Science of Our Planet; Dorling Kindersley Limited, 2021.

173. Steffen, W.; Richardson, K.; Rockström, J.; Cornell, S. E.; Fetzer, I.; Bennett, E. M.; Biggs, R.; Carpenter, S. R.; De Vries, W.; De Wit, C. A.; Folke, C.; Gerten, D.; Heinke, J.; Mace, G. M.; Persson, L. M.; Ramanathan, V.; Reyers, B.; Sörlin, S. Planetary Boundaries: Guiding Human Development on a Changing Planet. Science 2015, 347, No. 1259855.

174. Rockström, J.; Steffen, W.; Noone, K.; Persson, Å.; Chapin, F. S.; Lambin, E. F.; Lenton, T. M.; Scheffer, M.; Folke, C.; Schellnhuber, H. J.; Nykvist, B.; de Wit, C. A.; Hughes, T.; van der Leeuw, S.; Rodhe, H.; Sörlin, S.; Snyder, P. K.; Costanza, R.; Svedin, U.; Falkenmark, M.; Karlberg, L.; Corell, R. W.; Fabry, V. J.; Hansen, J.; Walker, B.; Liverman, D.; Richardson, K.; Crutzen, P.; Foley, J. A. A Safe Operating Space for Humanity. Nature 2009, 461, 472–475.

175. Christensen, B.; Brentrup, F.; Six, L.; Palliere, C.; Hoxha, A. In Assesing the Carbon Footprint of Fertilizers at Production and Full Life Cycle, Proceedings 751; International Fertiliser Society: London, U.K., 2014; Vol. 751, p 20.

176. Galloway, J. N.; Aber, J. D.; Erisman, J. W.; Seitzinger, S. P.; Howarth, R. W.; Cowling, E. B.; Cosby, B. J. The Nitrogen Cascade. Bioscience 2003, 53, 341–356.

177. FAO StatisticsFood and Agriculture Organization of the United Nations; FAO: Rome, Italy, 2020.

178. Daneshgar, S.; Callegari, A.; Capodaglio, A. G.; Vaccari, D. The Potential Phosphorus Crisis: Resource Conservation and Possible Escape Technologies: A Review. Resources 2018, 7, No. 22.

179. Desmidt, E.; Ghyselbrecht, K.; Zhang, Y.; Pinoy, L.; Van Der Bruggen, B.; Verstraete, W.; Rabaey, K.; Meesschaert, B. Global Phosphorus Scarcity and

Full-Scale P-Recovery Techniques: A Review. Crit. Rev. Environ. Sci. Technol. 2015, 45, 336–384.

180. Stahel, W. R. The Circular Economy: A User's Guide, 1st ed.; MacArthur, E. F., Ed.; Taylor & Francis: New York, NY, 2019.

181. Pigoli, A.; Zilio, M.; Tambone, F.; Mazzini, S.; Schepis, M.; Meers, E.; Schoumans, O.; Giordano, A.; Adani, F. Thermophilic Anaerobic Digestion as Suitable Bioprocess Producing Organic and Chemical Renewable Fertilizers: A Full-Scale Approach. Waste Manage. 2021, 124, 356–367.

182. McDonough, W.; Braungart, M. Cradle to Cradle: Remaking the Way We Make Things, 1st ed.; North Point Press: New York, 2002.

183. Tambone, F.; Scaglia, B.; D'Imporzano, G.; Schievano, A.; Orzi, V.; Salati, S.; Adani, F. Assessing Amendment and Fertilizing Properties of Digestates from Anaerobic Digestion through a Comparative Study with Digested Sludge and Compost. Chemosphere 2010, 81, 577–583.

184. Yasar, A.; Rasheed, R.; Tabinda, A. B.; Tahir, A.; Sarwar, F. Life Cycle Assessment of a Medium Commercial Scale Biogas Plant and Nutritional Assessment of Effluent Slurry. Renewable Sustainable Energy Rev. 2017, 67, 364–371.

185. Mazzini, S.; Borgonovo, G.; Scaglioni, L.; Bedussi, F.; D'Imporzano, G.; Tambone, F.; Adani, F. Phosphorus Speciation during Anaerobic Digestion and Subsequent Solid/Liquid Separation. Sci. Total Environ. 2020, 734, No. 139284.

186. Sigurnjak, I.; Brienza, C.; Snauwaert, E.; De Dobbelaere, A.; De Mey, J.; Vaneeckhaute, C.; Michels, E.; Schoumans, O.; Adani, F.; Meers, E. Production and Performance of Bio-Based Mineral Fertilizers from

Agricultural Waste Using Ammonia (Stripping-)Scrubbing Technology. Waste Manage. 2019, 89, 265–274.

187. Ledda, C.; Schievano, A.; Salati, S.; Adani, F. Nitrogen and Water Recovery from Animal Slurries by a New Integrated Ultrafiltration, Reverse Osmosis and Cold Stripping Process: A Case Study. Water Res. 2013, 47, 6157–6166.

188. Pepè Sciarria, T.; Vacca, G.; Tambone, F.; Trombino, L.; Adani, F. Nutrient Recovery and Energy Production from Digestate Using Microbial Electrochemical Technologies (METs). J. Cleaner Prod. 2019, 208, 1022–1029.

189. Sutton, M. A.; Bleeker, A.; Howard, C. M.; Bekunda, M.; Grizzetti, B.; de Vries, W.; van Grinsven, H. J. M.; Abrol, Y. P.; Adhya, T. K.; Billen, G.; Davidson, E.; Datta, A.; Diaz, R.; Erisman, J. W.; Liu, X. J.; Oenema, O.; Palm, C.; Raghuram, N.; Reis, S.; Scholz, R. W.; Sims, T.; Westhoek, H.; Zhang, F. S. Our Nutrient World: The Challenge to Produce More Food and Energy with Less Pollution. Global Overview of Nutrient Management; Centre for Ecology & Hydrology, 2018; Vol. 47.

190. Anastasiou, D.; Aran, M.; Balesdent, J.; Basch, G.; Biró, B.; de Maria Mourao, I.; Costantini, E.; Dell̃ 'Abate, M. T.; Dietz, S.; Follain, S.; Gomez-Macpherson, H.; Konsten, C.; Lloveras, J.; Marques, F.; Clara Martínez Gaitán, C.; Mavridis, A.; Neeteson, J.; Perdigao, A.;̃ Sarno, G.; Theocharopoulos, S.; Blanco, J.; Ambar, M.; Hinsinger, P. Soil Organic Matter in Mediterranean Regions; Brussels, Belgium, 2015.

191. Lal, R. Challenges and Opportunities in Soil Organic Matter Research. Eur. J. Soil Sci. 2009, 60, 158–169.

192. Zhang, X.; Davidson, E. A.; Mauzerall, D. L.; Searchinger, T. D.; Dumas, P.; Shen, Y. Managing Nitrogen for Sustainable Development. Nature 2015, 528, 51–59.

193. Riva, C.; Orzi, V.; Carozzi, M.; Acutis, M.; Boccasile, G.; Lonati, S.; Tambone, F.; D'Imporzano, G.; Adani, F. Short-Term Experiments in Using Digestate Products as Substitutes for Mineral (N) Fertilizer: Agronomic Performance, Odours, and Ammonia Emission Impacts. Sci. Total Environ. 2016, 547, 206–214.

194. Zilio, M.; Pigoli, A.; Rizzi, B.; Geromel, G.; Meers, E.; Schoumans, O.; Giordano, A.; Adani, F. Measuring Ammonia and Odours Emissions during Full Field Digestate Use in Agriculture. Sci. Total Environ. 2021, 782, No. 146882.

195. Webb, J.; Pain, B.; Bittman, S.; Morgan, J. The Impacts of Manure Application Methods on Emissions of Ammonia, Nitrous Oxide and on Crop Response-A Review. Agric., Ecosyst. Environ. 2010, 137, 39–46.

196. Basosi, R.; Spinelli, D.; Fierro, A.; Jez, S. Mineral Nitrogen Fertilizers: Environmental Impact of Production and Use. In Fertilizers Components, Uses in Agriculture and Environmental Impacts;López-Valdez, F.; Fernández-Luqueño, F., Eds.; Nova Science Publisher, Inc.: New York, NY, 2014; p 316.

197. Webb, J.; Sørensen, P.; Velthof, G.; Amon, B.; Pinto, M.; Rodhe, L.; Salomon, E.; Hutchings, N.; Burczyk, P.; Reid, J. An Assessment of the Variation of Manure Nitrogen Efficiency throughout Europe and an Appraisal of Means to Increase Manure N Efficiency. In Advances in Agronomy; Elsevier, 2013; Vol. 119, pp 371–442.

198. Styles, D.; Adams, P.; Thelin, G.; Vaneeckhaute, C.; Chadwick, D.; Withers, P. J. A. Life Cycle Assessment of Biofertilizer Production and Use Compared with Conventional Liquid Digestate Management. Environ. Sci. Technol. 2018, 52, 7468−7476.

199. Hijazi, O.; Munro, S.; Zerhusen, B.; Effenberger, M. Review of Life Cycle Assessment for Biogas Production in Europe. Renewable Sustainable Energy Rev. 2016, 54, 1291−1300.

200. Florio, C.; Fiorentino, G.; Corcelli, F.; Ulgiati, S.; Dumontet, S.; Güsewell, J.; Eltrop, L. A Life Cycle Assessment of Biomethane Production from Waste Feedstock through Different Upgrading Technologies. Energies 2019, 12, No. 718.

201. Li, J.; Xiong, F.; Chen, Z. An Integrated Life Cycle and Water Footprint Assessment of Nonfood Crops Based Bioenergy Produc tion. Sci. Rep. 2021, 11, No. 3912.

202. Timonen, K.; Sinkko, T.; Luostarinen, S.; Tampio, E.; Joensuu, K. LCA of Anaerobic Digestion: Emission Allocation for Energy and Digestate. J. Cleaner Prod. 2019, 235, 1567−1579.

203. Montemayor, E.; Bonmatí, A.; Torrellas, M.; Camps, F.; Ortiz, C.; Domingo, F.; Riau, V.; Antón, A. Environmental Accounting of Closed-Loop Maize Production Scenarios: Manure as Fertilizer and Inclusion of Catch Crops. Resour., Conserv. Recycl. 2019, 146, 395− 404.

204. Lyng, K. A.; Modahl, I. S.; Møller, H.; Saxegård, S. Comparison of Results from Life Cycle Assessment When Using Predicted and Real-Life Data for an Anaerobic Digestion Plant. J. Sustainable Dev. Energy, Water Environ. Syst. 2021, 9, 1−14.

205. Klöpffer, W. The Critical Review of Life Cycle Assessment Studies According to ISO 14040 and 14044. Int. J. Life Cycle Assess. 2012, 17, 1087–1093.

206. Di Capua, F.; Adani, F.; Pirozzi, F.; Esposito, G.; Giordano, A. Air Side-Stream Ammonia Stripping in a Thin Film Evaporator Coupled to High-Solid Anaerobic Digestion of Sewage Sludge: Process Performance and Interactions. J. Environ. Manage. 2021, 295, No. 113075.

207. Regione Lombardia. Programma d'Azione Regionale per La Protezione Delle Acque Dall' Inquinamento Provocato Dai Nitrati Provenienti Da Fonti Agricole Nelle Zone Vulnerabili Ai Sensi Della Direttiva Nitrati 91/676/CEE, 2020.

208. Bertora, C.; Peyron, M.; Pelissetti, S.; Grignani, C.; Sacco, D. Assessment of Methane and Nitrous Oxide Fluxes from Paddy Field by Means of Static Closed Chambers Maintaining Plants within Headspace. J. Visualized Exp. 2018, 139, 1–7.

209. Piccini, I.; Arnieri, F.; Caprio, E.; Nervo, B.; Pelissetti, S.; Palestrini, C.; Roslin, T.; Rolando, A. Greenhouse Gas Emissions from Dung Pats Vary with Dung Beetle Species and with Assemblage Composition. PLoS One 2017, 12, No. e0178077.

210. Peyron, M.; Bertora, C.; Pelissetti, S.; Said-Pullicino, D.; Celi, L.; Miniotti, E.; Romani, M.; Sacco, D. Greenhouse Gas Emissions as Affected by Different Water Management Practices in Temperate Rice Paddies. Agric., Ecosyst. Environ. 2016, 232, 17–28.

211. Tang, Y. S.; Cape, J. N.; Sutton, M. A. Development and Types of Passive Samplers for Monitoring Atmospheric NO2 and NH3 Concentrations. Sci. World J. 2001, 1, 513–529.

212. Weidema, B. P.; Bauer, C.; Hischier, R.; Mutel, C.; Nemecek, T.; Reinhard, J.; Vadenbo, C. O.; Wernet, G. Overview and Methodology. Data Quality Guideline for the Ecoinvent Database Version 3; The ecoinvent Centre: St. Gallen, 2013; Vol. 3.

213. IPCC (Intergovernmental Panel on Climate Change). Guidelines for National Greenhouse Gas Inventories: Agriculture, Forestry and Other Land Use; IPCC: Geneva, Switzerland, 2006.

214. Nemecek, T.; Kägi, T. Life Cycle Inventories of Agricultural Production Systems. Ecoinvent Report N.15, Swiss Centre for Life Cycle Inventories; Zurig and Dubendorf: 2007, pp 46.

215. Xu, Y.; Yu, W.; Ma, Q.; Zhou, H. Accumulation of Copper and Zinc in Soil and Plant within Ten-Year Application of Different Pig Manure Rates. Plant, Soil Environ. 2013, 59, 492–499.

216. Börjesson, G.; Kirchmann, H.; Kätterer, T. Four Swedish Long Term Field Experiments with Sewage Sludge Reveal a Limited Effect on Soil Microbes and on Metal Uptake by Crops. J. Soils Sediments 2014, 14, 164–177.

217. Goedkoop, M.; De Schryver, A.; Oele, M.; Durksz, S.; de Roest, D. Introduction to LCA with SimaPro 7; PRéConsult.: Netherlands, 2008.

218. Huijbregts, M. A. J.; Steinmann, Z. J. N.; Elshout, P. M. F.; Stam, G.; Verones, F.; Vieira, M.; Zijp, M.; Hollander, A.; van Zelm, R. ReCiPe2016: A Harmonised Life Cycle Impact Assessment Method at Midpoint and Endpoint Level. Int. J. Life Cycle Assess. 2017, 22, 138–147.

219. LCA Compendium - The Complete World of Life Cycle Assessment; 1st ed.; Hauschild, M. Z.; Huijbregts, M. A. J., Eds.; Springer: Dordrecht Heidelberg New York London, 2015.

220. Burmaster, D. E.; Anderson, P. D. Principles of Good Practice for the Use of Monte Carlo Techniques in Human Health and Ecological Risk Assessments. Risk Anal. 1994, 14, 477–481.

221. Paolini, V.; Petracchini, F.; Segreto, M.; Tomassetti, L.; Naja, N.; Cecinato, A. Environmental Impact of Biogas: A Short Review of Current Knowledge. J. Environ. Sci. Health, Part A 2018, 53, 899–906.

222. Scaglia, B.; Tambone, F.; Corno, L.; Orzi, V.; Lazzarini, Y.; Garuti, G.; Adani, F. Potential Agronomic and Environmental Properties of Thermophilic Anaerobically Digested Municipal Sewage Sludge Measured by an Unsupervised and a Supervised Chemometric Approach. Sci. Total Environ. 2018, 637–638, 791–802.

223. Tambone, F.; Adani, F. Nitrogen Mineralization from Digestate in Comparison to Sewage Sludge, Compost and Urea in a Laboratory Incubated Soil Experiment. J. Plant Nutr. Soil Sci. 2017, 180, 355– 365.

224. Brentrup, F.; Kusters, J.; Lammel, J.; Kuhlmann, H. Methods to Estimate On-Field Nitrogen Emissions from Crop Production as an Input to LCA Studies in the Agricultural Sector. Int. J. Life Cycle Assess. 2000, 5, No. 349.

225. Emmerling, C.; Krein, A.; Junk, J. Meta-Analysis of Strategies to Reduce NH3 Emissions from Slurries in European Agriculture and Consequences for Greenhouse Gas Emissions. Agronomy 2020, 10, No. 1633.

226. Björnsson, L.; Lantz, M.; Börjesson, P.; Prade, T.; Svensson, S.- E.; Eriksson, H. Impact of Biogas Crop Production on Greenhouse Gas Emissions, Soil Organic Matter and Food Crop Production—A Case Study on Farm Level; The Swedish Knowledge Centre for Renewable Transportation Fuels, 2013.

227. Willén, A.; Junestedt, C.; Rodhe, L.; Pell, M.; Jönsson, H. Sewage Sludge as Fertiliser - Environmental Assessment of Storage and Land Application Options. Water Sci. Technol. 2017, 75, 1034– 1050.

228. Bacenetti, J.; Lovarelli, D.; Fiala, M. Mechanisation of Organic Fertiliser Spreading, Choice of Fertiliser and Crop Residue Manage ment as Solutions for Maize Environmental Impact Mitigation. Eur. J. Agron. 2016, 79, 107–118.

229. Macura, B.; Johannesdottir, S. L.; Piniewski, M.; Haddaway, N. R.; Kvarnström, E. Effectiveness of Ecotechnologies for Recovery of Nitrogen and Phosphorus from Anaerobic Digestate and Effectiveness of the Recovery Products as Fertilisers: A Systematic Review Protocol. Environ. Evidence 2019, 8, No. 29

230. Peccia, J.; Westerhoff, P. We Should Expect More out of Our Sewage Sludge. Environ. Sci. Technol. 2015, 49, 8271–8276.

231. Provolo, G.; Manuli, G.; Finzi, A.; Lucchini, G.; Riva, E.; Sacchi, G. A. Effect of Pig and Cattle Slurry Application on Heavy Metal Composition of Maize Grown on Different Soils. Sustainability 2018, 10, No. 2684.

232. Leclerc, A.; Laurent, A. Framework for Estimating Toxic Releases from the Application of Manure on Agricultural Soil: National Release Inventories for Heavy Metals in 2000–2014. Sci. Total Environ. 2017, 590–591, 452–460.

233. Ertl, K.; Goessler, W. Grains, Whole Flour, White Flour, and Some Final Goods: An Elemental Comparison. Eur. Food Res. Technol. 2018, 244, 2065–2075

234. Niero, M.; Pizzol, M.; Bruun, H. G.; Thomsen, M. Comparative Life Cycle Assessment of Wastewater Treatment in Denmark Including Sensitivity and Uncertainty Analysis. J. Cleaner Prod. 2014, 68, 25–35.

235. Bacenetti, J.; Sala, C.; Fusi, A.; Fiala, M. Agricultural Anaerobic Digestion Plants: What LCA Studies Pointed out and What Can Be Done to Make Them More Environmentally Sustainable. Appl. Energy 2016, 179, 669–686.

236. Piippo, S.; Lauronen, M.; Postila, H. Greenhouse Gas Emissions from Different Sewage Sludge Treatment Methods in North. J. Cleaner Prod. 2018, 177, 483–492.

237. Yoshida, H.; ten Hoeve, M.; Christensen, T. H.; Bruun, S.; Jensen, L. S.; Scheutz, C. Life Cycle Assessment of Sewage Sludge Management Options Including Long-Term Impacts after Land Application. J. Cleaner Prod. 2018, 174, 538–547.

238. Lam, K. L.; Zlatanovic, L.; van der Hoek, J. P. Life Cycle Assessment of Nutrient Recycling from Wastewater: A Critical Review. Water Res. 2020, 173, No. 115519.

239. Menéndez, S.; Barrena, I.; Setien, I.; González-Murua, C.; Estavillo, J. M. Efficiency of Nitrification Inhibitor DMPP to Reduce Nitrous Oxide Emissions under Different Temperature and Moisture Conditions. Soil Biol. Biochem. 2012, 53, 82–89.

240. Herr, C.; Mannheim, T.; Müller, T.; Ruser, R. Effect of Nitrification Inhibitors on N2O Emissions after Cattle Slurry Application. Agronomy 2020, 10, No. 1174.

241. Qiao, C.; Liu, L.; Hu, S.; Compton, J. E.; Greaver, T. L.; Li, Q. How Inhibiting Nitrification Affects Nitrogen Cycle and Reduces Environmental Impacts of Anthropogenic Nitrogen Input. Global Change Biol. 2015, 21, 1249–1257.

242. Scarlat, N.; Dallemand, J.-F.; Fahl, F. Biogas: Developments and Perspectives in Europe. Renewable Energy 2018, 129, 457–472.

243. Benato, A.; Macor, A. Italian Biogas Plants: Trend, Subsidies, Cost, Biogas Composition and Engine Emissions. Energies 2019, 12, No. 979.

244. GSE (Gestore dei Servizi Energetici). Atlaimpianti di produzione di energia elettrica. https://atla.gse.it/atlaimpianti/ project/Atlaimpianti_Internet.html.

245. EBA; GIE; NGVA; SEA-LNG. BioLNG in Transport: Making Climate Neutrality a Reality A Joint White Paper about BioLNG Production; Brussels, Belgium, 2020.

246. Smajla, I.; Sedlar, D. K.; Drljaca, B.; Jukic´, L. Fuel Switch to´ LNG in Heavy Truck Traffic. Energies 2019, 12, No. 515.

247. Ali, Noureddine Shawky, Hamdallah Suleiman Rahi, and Abdel Wahab Abdel Razzaq Shaker. 2014. Soil fertility. Scientific Books House for Printing and Publishing.(in Arabic)

248. Peter, S. 2015. Managing Nitrogen in Crop Production. American Society of Agronomy, Inc. Crop Science Society of America, Inc. Soil Science Society of America, Inc.

249. Al-Hilli, Munther Majid Tajuddin. 2007. Efficiency of sulfur-coated urea in potassium release, phosphorous availability, and wheat yield and growth. PhD thesis - Department of Soil and Water Sciences - College of Agriculture - University of Baghdad.(in Arabic)

250. Havlin,J.L.; J.D. Beaton; S.L. Tisdale ; W.L. Nelson. 2005. Soil fertility and fertilizers, An Introduction to Nutrient Management,7thed, Upper Saddle River New Jersey. USA. p.515.

251. Al-Zubaidi, Bashar Mazher Jader. 2010. The effect of organic and potassium fertilizers on potassium availability and on growth and yield of yellow corn (Zea mays L.). Master Thesis. College of Agriculture - University of Baghdad - Iraq. (in Arabic)

252. Dadhich, S.K.,L.L. Somani, and D. shilpkar. 2011. Effect of integrated use of fertilizer p, FYM and bio fertilizers on soil properties and productivity of soybean- wheat crop sequence. Journal of Advances in Developmental Research. 2(1):42-46.

253. Ali, Noureddin Shawghi and Abdullah Abdul Razzaq Shaker. 2018. Organic fertilization and its role in sustainable agriculture. Scientific Books House - Al-Mutanabi - Baghdad - Iraq.(in Arabic)

254. Ali, N.S. and S.M.A. AL Khalel. 2016. Integrated Effect of Mineral, Organic and Bio Fertilization on the Productivity of Plastic House Tomato-A Transition Step towards Organic Farming IJRDO. J. Agric. Res. 2(1):1-6.

255. Mansour, Montazer Hammadi. 2014. The effect of phosphate, organic and bio fertilization on phosphorous readiness, growth and yield of yellow corn (Zea mays L.). Master Thesis. College of Agriculture / University of Baghdad. (in Arabic)

256. Mohammed, M.A. (2020). Structural, Optical, Electrical and Gas Sensor Properties of ZrO2 Thin Films prepared by Sol-Gel Technique. Neuroquantology, 18(3), 22-27. doi: 10.14704/nq.2020.18.3.nq20146

257. Ali, Muhammad Jasim and Mohsen Awaid Farhan (2012). Estimating cost functions and economies of scale for yellow maize (Babilon Governorate, an applied model) Iraqi Journal of Agricultural Sciences, 43 (2): 65-74.(in Arabic)

258. FAO, (Food and Agriculture Organization) (2012). Food and Agriculture Organization of the united nations. Rome. [13] Black, C.A. 1965 a. Methods of soil analysis. Part1. Physical and Mineralogical properties Am. Soc. Agron., 9.Madison,Wisconsin,USA

259. Jackson, M. L. (1958). Soil chemical analysis . Prentice – hall Inc. Englewood cliff, N.J.

260. Lanyon, L.E. and W.R. Heald. 1982. Magnesium, Calcium,Strontium and Barium in A.L Page , (Ed). Methods of Soil Analysis. Part2. Chemical and Microbiological Properties 2nd edition, Amer. Soc. of Agron. Inc. Soil Sci. Soc. Am. Inc. Madision. Wis. U.S.A.

261. Raid SH. Jarallah and Nihad A. Abbas. (2019). The Effect of Sulfur and Phosphate Fertilizers Application on the Dissolved Phosphorus Amount in Rhizosphere of Zea Maize L.". Al-Qadisiyah Journal For Agriculture Sciences, 9(2), 233-239.

262. Page, A.I.; R.H. Miller and D.R. Kenney. 1982. Methods of Soil Analysis, Part (2). 2nd ed. Agronomy 9 Am. Soc. Agron. Madison, Wisconsin.

263. Black, C.A. 1965 b. Methods of Soil Analysis. Part2. Chemical and microbiological properties Am. Soc. Agron. , Inc. Madison Wisconson, USA

264. Al-Sahhaf, Fadel Hussain (1989). Applied plant nutrition. House of wisdom. Ministry of Higher Education and Scientific Research. Iraq.(in Arabic)

265. Schaffelen, A.C.A. and Vanschauwenbury,J.C.H.(1960).Quick test for soil and plant analysis used by small laboratories .Neth.J.Agric.Sci.9:2-16.

266. Batista B.D., Lacava P.T., Ferrari A., Teixeira-Silva N.S., Bonatelli M.L., Tsui S., Mondin M., Kitajima E.W., Pereira J.O., Azevedo J.L. Screening of tropically derived, multi-trait plant growth-promoting rhizobacteria and evaluation of corn and soybean colonization ability. *Microbiol. Res.* 2018;206:33–42. doi: 10.1016/j.micres.2017.09.007. [PubMed] [CrossRef] [Google Scholar]

267. Kumar S., Reddy C., Phogat M., Korav S. Role of bio-fertilizers towards sustainable agricultural development: A review. *J. Pharm. Phytochem.* 2018;7:1915–1921. [Google Scholar]

268. Glick B.R. Bacteria with ACC deaminase can promote plant growth and help to feed the world. *Microbiol. Res.* 2014;169:30–39. doi: 10.1016/j.micres.2013.09.009. [PubMed] [CrossRef] [Google Scholar]

269. 4. Debnath K., Conway G. *One Billion Hungry: Can We Feed the World?* Cornell University Press; Ithaca, NY, USA: 2012. p. 456. [Google Scholar]

270. Mahanty T., Bhattacharjee S., Goswami M., Bhattacharyya P., Das B., Ghosh A., Tribedi P. Biofertilizers: A potential approach for sustainable agriculture development. *Environ. Sci. Pollut. Res.* 2017;24:3315–3335. doi: 10.1007/s11356-016-8104-0. [PubMed] [CrossRef] [Google Scholar]

271. Egamberdieva D., Shrivastava S., Varma A. *Plant-Growth-Promoting Rhizobacteria (PGPR) and Medicinal Plants.* Springer; Berlin/Heidelberg, Germany: 2015. [Google Scholar]

272. Liu Y., Pan X., Li J. Current agricultural practices threaten future global food production. *J. Agric. Environ. Ethics.* 2015;28:203–216. doi: 10.1007/s10806-014-9527-6. [CrossRef] [Google Scholar]

273. Bruinsma J. The resource outlook to 2050: By how much do land, water and crop yields need to increase by 2050?; Proceedings of the How to Feed the World in 2050, Proceedings of a Technical Meeting of Experts; Rome, Italy. 24–26 June 2009; pp. 1–33. [Google Scholar]

274. Sinha R.K. The concept of sustainable agriculture: An issue of food safety & security for people, economic prosperity for the farmers and ecological

security for the nations. *Am. Eurasian J. Agric. Environ. Sci.* 2009;5:1–4. [Google Scholar]

275. Wang C.L., Chien S.Y., Young C.C. Present situation and future perspective of bio-fertilizer for environmentally friendly agriculture. *Annu. Rep.* 2014:1–5. [Google Scholar]

276. Youssef M., Eissa M. Biofertilizers and their role in management of plant parasitic nematodes. *A review. J. Biotechnol. Pharm. Res.* 2014;5:1–6. [Google Scholar]

277. Malusa E., Vassilev N. A contribution to set a legal framework for biofertilisers. *Appl. Microbiol. Biotechnol.* 2014;98:6599–6607. doi: 10.1007/s00253-014-5828-y. [PMC free article] [PubMed] [CrossRef] [Google Scholar]

278. Bardi L., Malusà E. *Abiotic Stress: New Research.* Nova Science Publishers, Inc.; Hauppauge, NY, USA: 2012. Drought and nutritional stresses in plant: Alleviating role of rhizospheric microorganisms; pp. 1–57. [Google Scholar]

279. Mazid M., Khan T.A. Future of bio-fertilizers in Indian agriculture: An overview. *Int. J. Agric. Food Res.* 2015;3:10–23. doi: 10.24102/ijafr.v3i3.132. [CrossRef] [Google Scholar]

280. Raja N. Biopesticides and biofertilizers: Ecofriendly sources for sustainable agriculture. *J. Biofertil. Biopestic.* 2013;4:1–2. doi: 10.4172/2155-6202.1000e112. [CrossRef] [Google Scholar]

281. Stewart W., Roberts T. Food security and the role of fertilizer in supporting it. *Procedia Eng.* 2012;46:76–82. doi: 10.1016/j.proeng.2012.09.448. [CrossRef] [Google Scholar]

282. Nosheen S., Ajmal I., Song Y. Microbes as biofertilizers, a potential approach for sustainable crop production. *Sustainability.* 2021;13:1868. doi: 10.3390/su13041868. [CrossRef] [Google Scholar]

283. Bumandalai O., Tserennadmid R. Effect of *Chlorella vulgaris* as a biofertilizer on germination of tomato and cucumber seeds. *Int. J. Aquat. Biol.* 2019;7:95–99. [Google Scholar]

284. Umesha S., Singh P.K., Singh R.P. *Biotechnology for Sustainable Agriculture.* Elsevier; Amsterdam, The Netherlands: 2018. Microbial biotechnology and sustainable agriculture; pp. 185–205. [Google Scholar]

285. Parikh S.J., James B.R. Soil: The foundation of agriculture. *Nat. Educ. Knowl.* 2012;3:2. [Google Scholar]

286. Raynaud X., Nunan N. Spatial ecology of bacteria at the microscale in soil. *PLoS ONE.* 2014;9:e87217. doi: 10.1371/journal.pone.0087217. [PMC free article] [PubMed] [CrossRef] [Google Scholar]

287. Etesami H., Emami S., Alikhani H.A. Potassium solubilizing bacteria (KSB): Mechanisms, promotion of plant growth, and future prospects: A review. *J. Soil Sci. Plant Nutr.* 2017;17:897–911. doi: 10.4067/S0718-95162017000400005. [CrossRef] [Google Scholar]

288. Jha Y. Potassium mobilizing bacteria: Enhance potassium intake in paddy to regulates membrane permeability and accumulate carbohydrates under salinity stress. *Braz. J. Biol. Sci.* 2017;4:333–344. doi: 10.21472/bjbs.040812. [CrossRef] [Google Scholar]

289. Itelima J., Bang W., Onyimba I., Sila M., Egbere O. Bio-fertilizers as key player in enhancing soil fertility and crop productivity: A review. *Direct Res. J. Agric. Food Sci.* 2018;6:73–83. [Google Scholar]

290. Kamran S., Shahid I., Baig D.N., Rizwan M., Malik K.A., Mehnaz S. Contribution of zinc solubilizing bacteria in growth promotion and zinc content of wheat. *Front. Microbiol.* 2017;8:2593. doi: 10.3389/fmicb.2017.02593. [PMC free article] [PubMed] [CrossRef] [Google Scholar]

291. Gouda S., Kerry R.G., Das G., Paramithiotis S., Shin H.-S., Patra J.K. Revitalization of plant growth promoting rhizobacteria for sustainable development in agriculture. *Microbiol. Res.* 2018;206:131–140. doi: 10.1016/j.micres.2017.08.016. [PubMed] [CrossRef] [Google Scholar]

292. Kurepin L.V., Zaman M., Pharis R.P. Phytohormonal basis for the plant growth promoting action of naturally occurring biostimulators. *J. Sci. Food Agric.* 2014;94:1715–1722. doi: 10.1002/jsfa.6545. [PubMed] [CrossRef] [Google Scholar]

293. Gupta G., Parihar S.S., Ahirwar N.K., Snehi S.K., Singh V. Plant growth promoting rhizobacteria (PGPR): Current and future prospects for development of sustainable agriculture. *J. Microb. Biochem. Technol.* 2015;7:096–102. [Google Scholar]

294. Ahemad M., Khan M.S. Evaluation of plant-growth-promoting activities of rhizobacterium *Pseudomonas putida* under herbicide stress. *Ann. Microbiol.* 2012;62:1531–1540. doi: 10.1007/s13213-011-0407-2. [CrossRef] [Google Scholar]

295. Glick B.R. Plant growth-promoting bacteria: Mechanisms and applications. *Scientifica.* 2012;2012:963401. doi: 10.6064/2012/963401. [PMC free article] [PubMed] [CrossRef] [Google Scholar]

296. Jahanian A., Chaichi M., Rezaei K., Rezayazdi K., Khavazi K. The effect of plant growth promoting rhizobacteria (PGPR) on germination and primary growth of artichoke (*Cynara scolymus*) *Int. J. Agric. Crop Sci.* 2012;4:923–929. [Google Scholar]

297. Liu W., Wang Q., Hou J., Tu C., Luo Y., Christie P. Whole genome analysis of halotolerant and alkalotolerant plant growth-promoting rhizobacterium Klebsiella sp. D5A. *Sci. Rep.* 2016;6:26710. doi: 10.1038/srep26710. [PMC free article] [PubMed] [CrossRef] [Google Scholar]

298. Xie J., Shi H., Du Z., Wang T., Liu X., Chen S. Comparative genomic and functional analysis reveal conservation of plant growth promoting traits in *Paenibacillus polymyxa* and its closely related species. *Sci. Rep.* 2016;6:21329. doi: 10.1038/srep21329. [PMC free article] [PubMed] [CrossRef] [Google Scholar]

299. Egamberdieva D., Lugtenberg B. *Use of Microbes for the Alleviation of Soil Stresses.* Volume 1. Springer; Berlin/Heidelberg, Germany: 2014. Use of plant growth-promoting rhizobacteria to alleviate salinity stress in plants; pp. 73–96. [Google Scholar]

300. Kundan R., Pant G., Jadon N., Agrawal P. Plant growth promoting rhizobacteria: Mechanism and current prospective. *J. Fertil. Pestic.* 2015;6:9. doi: 10.4172/2471-2728.1000155. [CrossRef] [Google Scholar]

301. Tairo E.V., Ndakidemi P.A. Possible benefits of rhizobial inoculation and phosphorus supplementation on nutrition, growth and economic sustainability in grain legumes. *Am. J. Res. Commun.* 2013;1:532–556. [Google Scholar]

302. Kumaar S., Babu R.P., Vivek P., Saravanan D. Role of Nitrogen Fixers as Biofertilizers in Future Perspective: A Review. *Res. J. Pharm.*

Technol. 2020;13:2459–2467. doi: 10.5958/0974-360X.2020.00440.0. [CrossRef] [Google Scholar]

303. Bhattacharyya P.N., Jha D.K. Plant growth-promoting rhizobacteria (PGPR): Emergence in agriculture. *World J. Microbiol. Biotechnol.* 2012;28:1327–1350. doi: 10.1007/s11274-011-0979-9. [PubMed] [CrossRef] [Google Scholar]

304. Smith B.E., Richards R.L., Newton W.E. *Catalysts for Nitrogen Fixation: Nitrogenases, Relevant Chemical Models and Commercial Processes.* Volume 1 Springer Science & Business Media; Berlin/Heidelberg, Germany: 2004. [Google Scholar]

305. Seefeldt L.C., Hoffman B.M., Dean D.R. Mechanism of Mo-dependent nitrogenase. *Annu. Rev. Biochem.* 2009;78:701–722. doi: 10.1146/annurev.biochem.78.070907.103812. [PMC free article] [PubMed] [CrossRef] [Google Scholar]

306. Allito B.B., Nana E.-M., Alemneh A.A. Rhizobia strain and legume genome interaction effects on nitrogen fixation and yield of grain legume: A review. *Mol. Soil Biol.* 2015;6:1–6. doi: 10.5376/msb.2015.06.0004. [CrossRef] [Google Scholar]

307. Verma J., Yadav J., Tiwari K., Lavakush S., Singh V. Impact of plant growth promoting rhizobacteria on crop production. *Int. J. Agric. Res.* 2010;5:954–983. doi: 10.3923/ijar.2010.954.983. [CrossRef] [Google Scholar]

308. Santi C., Bogusz D., Franche C. Biological nitrogen fixation in non-legume plants. *Ann. Bot.* 2013;111:743–767. doi: 10.1093/aob/mct048. [PMC free article] [PubMed] [CrossRef] [Google Scholar]

309. Burns R.C., Hardy R.W. *Nitrogen Fixation in Bacteria and Higher Plants.* Springer; Berlin/Heidelberg, Germany: 2012. [Google Scholar]

310. 45. Ahemad M., Kibret M. Mechanisms and applications of plant growth promoting rhizobacteria: Current perspective. *J. King Saud Univ. Sci.* 2014;26:1–20. doi: 10.1016/j.jksus.2013.05.001. [CrossRef] [Google Scholar]

311. Shamseldin A. The role of different genes involved in symbiotic nitrogen fixation—Review. *Glob. J. Biotechnol. Biochem.* 2013;8:84–94. [Google Scholar]

312. Coppola D., Giordano D., Tinajero-Trejo M., Di Prisco G., Ascenzi P., Poole R.K., Verde C. Antarctic bacterial haemoglobin and its role in the protection against nitrogen reactive species. *Biochim. Biophys. Acta (BBA) Proteins Proteom.* 2013;1834:1923–1931. doi: 10.1016/j.bbapap.2013.02.018. [PubMed] [CrossRef] [Google Scholar]

313. Glick B.R. *Beneficial Plant-Bacterial Interactions.* Springer; Berlin/Heidelberg, Germany: 2015. [Google Scholar]

314. Suzaki T., Yoro E., Kawaguchi M. Leguminous plants: Inventors of root nodules to accommodate symbiotic bacteria. *Int. Rev. Cell Mol. Biol.* 2015;316:111–158. [PubMed] [Google Scholar]

315. Suzaki T., Kawaguchi M. Root nodulation: A developmental program involving cell fate conversion triggered by symbiotic bacterial infection. *Curr. Opin. Plant Biol.* 2014;21:16–22. doi: 10.1016/j.pbi.2014.06.002. [PubMed] [CrossRef] [Google Scholar]

316. Maillet F., Poinsot V., André O., Puech-Pagès V., Haouy A., Gueunier M., Cromer L., Giraudet D., Formey D., Niebel A. Fungal lipochitooligosaccharide symbiotic signals in arbuscular mycorrhiza. *Nature.* 2011;469:58–63. doi: 10.1038/nature09622. [PubMed] [CrossRef] [Google Scholar]

317. Vijayan R., Palaniappan P., Tongmin S., Elavarasi P., Manoharan N. Rhizobitoxine enhances nodulation by inhibiting ethylene synthesis of *Bradyrhizobium elkanii* from Lespedeza species: Validation by homology modelingand molecular docking study. *World J. Pharm. Pharm. Sci.* 2013;2:4079–4094. [Google Scholar]

318. Nascimento F.X., Brígido C., Glick B.R., Oliveira S. ACC deaminase genes are conserved among *Mesorhizobium* species able to nodulate the same host plant. *FEMS Microbiol. Lett.* 2012;336:26–37. doi: 10.1111/j.1574-6968.2012.02648.x. [PubMed] [CrossRef] [Google Scholar]

319. Zahir Z.A., Zafar-ul-Hye M., Sajjad S., Naveed M. Comparative effectiveness of *Pseudomonas* and *Serratia* sp. containing ACC-deaminase for coinoculation with *Rhizobium leguminosarum* to improve growth, nodulation, and yield of lentil. *Biol. Fertil. Soils.* 2011;47:457–465. doi: 10.1007/s00374-011-0551-7. [CrossRef] [Google Scholar]

320. Gourion B., Berrabah F., Ratet P., Stacey G. Rhizobium–legume symbioses: The crucial role of plant immunity. *Trends Plant Sci.* 2015;20:186–194. doi: 10.1016/j.tplants.2014.11.008. [PubMed] [CrossRef] [Google Scholar]

321. Mehnaz S. *Plant Microbes Symbiosis: Applied Facets.* Springer; Berlin/Heidelberg, Germany: 2015. Azospirillum: A biofertilizer for every crop; pp. 297–314. [Google Scholar]

322. Mishra D., Rajvir S., Mishra U., Kumar S.S. Role of bio-fertilizer in organic agriculture: A review. *Res. J. Recent Sci.* 2013;2277:2502. [Google Scholar]

323. Mishra P., Dash D. Rejuvenation of biofertilizer for sustainable agriculture and economic development. *Consilience.* 2014;11:41–61. [Google Scholar]

324. Trabelsi D., Mhamdi R. Microbial inoculants and their impact on soil microbial communities: A review. *BioMed Res. Int.* 2013;2013:863240.

doi: 10.1155/2013/863240. [PMC free article] [PubMed] [CrossRef] [Google Scholar]

325. Naiman A.D., Latrónico A., de Salamone I.E.G. Inoculation of wheat with *Azospirillum brasilense* and *Pseudomonas fluorescens*: Impact on the production and culturable rhizosphere microflora. *Eur. J. Soil Biol.* 2009;45:44–51. doi: 10.1016/j.ejsobi.2008.11.001. [CrossRef] [Google Scholar]

326. Moraditochaee M., Azarpour E., Bozorgi H.R. Study effects of bio-fertilizers, nitrogen fertilizer and farmyard manure on yield and physiochemical properties of soil in lentil farming. *Int. J. Biosci.* 2014;4:41–48. [Google Scholar]

327. Mathivanan R., Umavathi S., Ramasamy P., Thangam Y. Influence of vermicompost on the activity of the plant growth regulators in the leaves of the Indian butter bean plant, *Dolichos lab lab* L. *Int. J. Adv. Res. Biol. Sci.* 2015;2:84–89. [Google Scholar]

328. Martin X.M., Sumathi C.S., Kannan V.R. Influence of agrochemicals and *Azotobacter* sp. application on soil fertility in relation to maize growth under nursery conditions. *Eurasian J. Biosci.* 2011;5:19–28. doi: 10.5053/ejobios.2011.5.0.3. [CrossRef] [Google Scholar]

329. Wani S.A., Chand S., Ali T. Potential use of *Azotobacter chroococcum* in crop production: An overview. *Curr. Agric. Res. J.* 2013;1:35–38. doi: 10.12944/CARJ.1.1.04. [CrossRef] [Google Scholar]

330. Wagner S.C. Biological nitrogen fixation. *Nat. Educ. Knowl.* 2011;3:15. [Google Scholar]

331. 66. Al Abboud M., Ghany T.A., Alawlaqi M. Role of biofertilizers in agriculture: A brief review. *Mycopath.* 2014;11:95–101. [Google Scholar]

332. Hamid A., Ahmad L. Soil phosphorus fixation chemistry and role of phosphate solubilizing bacteria in enhancing its efficiency for sustainable cropping—A review. *J. Pure Appl. Microbiol.* 2012;66:1905–1911. [Google Scholar]

333. McComb R.B., Bowers G.N., Jr., Posen S. *Alkaline Phosphatase.* Springer Science & Business Media; Berlin/Heidelberg, Germany: 2013. [Google Scholar]

334. Pereira S.I., Castro P.M. Phosphate-solubilizing rhizobacteria enhance *Zea mays* growth in agricultural P-deficient soils. *Ecol. Eng.* 2014;73:526–535. doi: 10.1016/j.ecoleng.2014.09.060. [CrossRef] [Google Scholar]

335. 70. Mohammadi K., Sohrabi Y. Bacterial biofertilizers for sustainable crop production: A review. *ARPN J Agric Biol Sci.* 2012;7:307–316. [Google Scholar]

336. Parray J.A., Jan S., Kamili A.N., Qadri R.A., Egamberdieva D., Ahmad P. Current perspectives on plant growth-promoting rhizobacteria. *J. Plant Growth Regul.* 2016;35:877–902. doi: 10.1007/s00344-016-9583-4. [CrossRef] [Google Scholar]

337. Ahemad M., Khan M.S. Assessment of plant growth promoting activities of rhizobacterium *Pseudomonas putida* under insecticide-stress. *Microbiol. J.* 2011;1:54–64. doi: 10.3923/mj.2011.54.64. [CrossRef] [Google Scholar]

338. Rajkumar M., Sandhya S., Prasad M., Freitas H. Perspectives of plant-associated microbes in heavy metal phytoremediation. *Biotechnol. Adv.* 2012;30:1562–1574. doi: 10.1016/j.biotechadv.2012.04.011. [PubMed] [CrossRef] [Google Scholar]

339. Thomine S., Lanquar V. *Transporters and Pumps in Plant Signaling.* Springer; Berlin/Heidelberg, Germany: 2011. Iron transport and signaling in plants; pp. 99–131. [Google Scholar]

340. Ljung K. Auxin metabolism and homeostasis during plant development. *Development.* 2013;140:943–950. doi: 10.1242/dev.086363. [PubMed] [CrossRef] [Google Scholar]

341. Spaepen S., Vanderleyden J. Auxin and plant-microbe interactions. *Cold Spring Harb. Perspect. Biol.* 2011;3:a001438. doi: 10.1101/cshperspect.a001438. [PMC free article] [PubMed] [CrossRef] [Google Scholar]

342. Hermosa R., Viterbo A., Chet I., Monte E. Plant-beneficial effects of *Trichoderma* and of its genes. *Microbiology.* 2012;158:17–25. doi: 10.1099/mic.0.052274-0. [PubMed] [CrossRef] [Google Scholar]

343. Mazurier S., Corberand T., Lemanceau P., Raaijmakers J.M. Phenazine antibiotics produced by fluorescent pseudomonads contribute to natural soil suppressiveness to Fusarium wilt. *ISME J.* 2009;3:977–991. doi: 10.1038/ismej.2009.33. [PubMed] [CrossRef] [Google Scholar]

344. Pieterse C.M., Zamioudis C., Berendsen R.L., Weller D.M., Van Wees S.C., Bakker P.A. Induced systemic resistance by beneficial microbes. *Annu. Rev. Phytopathol.* 2014;52:347–375. doi: 10.1146/annurev-phyto-082712-102340. [PubMed] [CrossRef] [Google Scholar]

345. Maksimov I., Abizgil'Dina R., Pusenkova L. Plant growth promoting rhizobacteria as alternative to chemical crop protectors from pathogens. *Appl. Biochem. Microbiol.* 2011;47:333–345. doi: 10.1134/S0003683811040090. [PubMed] [CrossRef] [Google Scholar]

346. Beneduzi A., Ambrosini A., Passaglia L.M. Plant growth-promoting rhizobacteria (PGPR): Their potential as antagonists and biocontrol agents. *Genet. Mol. Biol.* 2012;35:1044–1051. doi: 10.1590/S1415-47572012000600020. [PMC free article] [PubMed] [CrossRef] [Google Scholar]

347. Zhang P.J., Broekgaarden C., Zheng S.J., Snoeren T.A., van Loon J.J., Gols R., Dicke M. Jasmonate and ethylene signaling mediate whitefly-induced interference with indirect plant defense in *Arabidopsis thaliana. New Phytol.* 2013;197:1291–1299. doi: 10.1111/nph.12106. [PubMed] [CrossRef] [Google Scholar]

348. Tilman D., Cassman K.G., Matson P.A., Naylor R., Polasky S. Agricultural sustainability and intensive production practices. *Nature.* 2002;418:671–677. doi: 10.1038/nature01014. [PubMed] [CrossRef] [Google Scholar]

349. Aktar M.W., Sengupta D., Chowdhury A. Impact of pesticides use in agriculture: Their benefits and hazards. *Interdiscip. Toxicol.* 2009;2:1. doi: 10.2478/v10102-009-0001-7. [PMC free article] [PubMed] [CrossRef] [Google Scholar]

350. Pretty J., Bharucha Z.P. Integrated pest management for sustainable intensification of agriculture in Asia and Africa. *Insects.* 2015;6:152–182. doi: 10.3390/insects6010152. [PMC free article] [PubMed] [CrossRef] [Google Scholar]

351. Bhardwaj D., Ansari M.W., Sahoo R.K., Tuteja N. Biofertilizers function as key player in sustainable agriculture by improving soil fertility, plant tolerance and crop productivity. *Microb. Cell Factories.* 2014;13:66. doi: 10.1186/1475-2859-13-66. [PMC free article] [PubMed] [CrossRef] [Google Scholar]

352. Gaur V. Biofertilizer–necessity for sustainability. *J. Adv. Dev.* 2010;1:8.

353. Tao, Y.; Qiu, X.W.; Zhou, Y.X.; Hu, J.L. The conflict between risk perception, social trust and farmers' organic fertilizer substitution behavior. *J. Agrotech. Econ.* **2022**, *325*, 49–64. [**Google Scholar**]

354. Yang, Y.R.; Luo, X.F. The impact of reduction and replacement policy on farmers' willingness to adopt the organic fertilizer substitution technology model: An empirical analysis based on the survey data of tea growers in Hubei. *J. Agrotech. Econ.* **2018**, *282*, 77–85. [**Google Scholar**]

355. Kong, F.B.; Guo, Q.L.; Pan, D. Evaluation on overfertilization and its spatial-temporal difference about major grain crops in China. *Econ. Geogr.* **2018**, *38*, 201–210+240. [**Google Scholar**]

356. Jiang, H.P.; Yao, J.; Jiang, L. Development thoughts and policy suggestions for China's food security in the new era. *Economist* **2020**, *01*, 110–118. [**Google Scholar**]

357. Zhang, W.L.; Wu, S.X.; Ji, H.J.; Kolbe, H. Estimation of agricultural non-point source pollution in China and the alleviating strategies. *Sci. Agric. Sin.* **2004**, *07*, 1008–1017. [**Google Scholar**]

358. Jin, S.Q.; Wu, Y. Is agricultural non-point sources the primary cause of water pollution: Based on the data of the Huai River Basin. *Chin. Rural. Econ.* **2014**, *09*, 71–81. [**Google Scholar**]

359. Li, Q.; Li, K. Impact of pest control trusteeship on technical efficiency: From the perspective of horizontal division efficiencyand vertical coordination efficiency. *Resour. Sci.* **2022**, *44*, 1964–1979. [**Google Scholar**] [**CrossRef**]

360. Su, S.Y.; Zhou, X.; Zhou, Y.X. Can organic fertilizer substitution increase farmers' income: A case study of vegetable growers in Shandong. *J. Arid. Land Resour. Environ.* **2022**, *36*, 24–31. [**Google Scholar**]

361. Xu, D.Y.; Li, H.Y.; Sun, Y.X.; Wu, G.; Wang, J.B.; Yuan, M.M.; Wang, P.X.; Zhang, X.M.; Shu, X.H. Combined application of different proportions of organic and inorganic fertilizers:Effects on rice yield and nitrogen use efficiency. *Chin. Agric. Sci. Bull.* **2022**, *38*, 1–5. [**Google Scholar**]

362. Tian, B.; Wang, Y.P. Study on farmers' willingness to utilize straw resources and its driving factors: A case study of Wuhan and Changsha. *Rural. Econ.* **2014**, *09*, 102–107. [**Google Scholar**]

363. Mi, S.H.; Huang, Z.H.; Zhu, Q.B.; Huang, L.L. Study on factors influencing farmers' adoption of low-carbon technologies. *Acta Agric. Zhejiangensis* **2014**, *26*, 797–804. [**Google Scholar**]

364. Fang, S.M.; Kong, X.Z. Analysis of the impact of farmers' endowments on the adoption of production technologies in protected areas: A case study of Shaanxi, Sichuan and Ningxia. *J. Agrotech. Econ.* **2005**, *03*, 35–42. [**Google Scholar**]

365. Cai, S.K. Empirical study on economic structure, land's feature and green pest control technology adoption: Based on the Anhui Province 740 rice farmers' research. *J. China Agric. Univ.* **2013**, *18*, 208–215. [**Google Scholar**]

366. Li, X.; Mu, Y.Y. On the sustainable production behavior of vegetable farmers in northern protected region. *China Popul. Resour. Environ.* **2013**, *23*, 164–169. [**Google Scholar**]

367. Song, B.; Mu, Y.Y.; Hou, L.L. Study on the effect of farm households' specialization on low-carbon agriculture: Evidence from vegetable growers in Beijing, China. *J. Nat. Resour.* **2016**, *31*, 468–476. [**Google Scholar**]

368. Huang, J.K.; Qi, L.; Chen, R.J. Technical information knowledge, risk appetite and pesticide application by farmers. *J. Manag. World* **2008**, *05*, 71–76. [**Google Scholar**]

369. Wossen, T.; Berger, T.; Mequaninte, T.; Alamirew, B. Social network effects on the adoption of sustainable natural resource management practices in Ethiopia. *Int. J. Sustain. Dev. World Ecol.* **2013**, *20*, 477–483. [**Google Scholar**] [**CrossRef**]

370. Qiao, D.; Lu, Q.; Xu, T. Social networks, information acquisition and water-saving irrigation technology adoption: An empirical analysis from Minqin County, Gansu Province. *J. Nanjing Agric. Univ. Soc. Sci. Ed.* **2017**, *17*, 147–155+160. [**Google Scholar**]

371. Cai, S.K.; Ni, P.F. Study on the yield-increasing effect of integrated rice pest and disease management technology: A case study of late rice. *Food Econ. Res.* **2015**, *1*, 65–75. [**Google Scholar**]

372. Liu, L.; Zhang, J.; Zhang, S.C.; Qiu, H.G. Will the expansion of operations help farmers adopt environmentally friendly production behaviors: A case study of straw returning. *J. Agrotech. Econ.* **2017**, *05*, 17–26. [**Google Scholar**]

373. Kurkalova, L.; Kling, C.; Zhao, J. Green Subsidies in Agriculture: Estimating the Adoption Costs of Conservation Tillage from Observed Behavior. *Can. J. Agric. Econ.* **2006**, *54*, 247–267. [**Google Scholar**] [**CrossRef**][**Green Version**]

374. Ren, X.F.; Li, X.P. Income maximization and behavior of cultivated land protection of farmer in China. *China Popul. Resour. Environ.* **2011**, *21*, 79–85. [**Google Scholar**]

375. Colette, W.A.; Almas, L.K.; Schuster, G.L. Evaluating the impact of integrated pest management on agriculture and the environment in the Texas panhandle. *Gen. Inf.* **2001**, *26*, 561–562. [**Google Scholar**]

376. Cai, R.; Cai, S.K. The adoption of conservation agricultural technology and the impact on crop yields based on rice farms in Anhui Province. *Resour. Sci.* **2012**, *34*, 1705–1711. [**Google Scholar**]

377. Dong, Y.; Mu, Y.Y. The path selection and efficiency increase mechanism of farmers' adoption of environmentally friendly technology: An empirical analysis. *China Rural. Surv.* **2019**, *02*, 34–48. [**Google Scholar**]

378. Wang, J.X.; Zhang, L.J. Impacts of conservation tillage on agriculture: Empirical research in the Yellow River Basin. *Manag. Rev.* **2010**, *22*, 77–84+60. [**Google Scholar**]

379. Tao, R.; Xu, Z.G.; Xu, J.T. Returning farmland to forests, food policy and sustainable development. *Soc. Sci. China* **2004**, *6*, 25–38+204. [**Google Scholar**]

380. Xu, Q.; Zhang, Y. Land adjustment, land rights stability and long-term investment incentives for farmers. *Econ. Res. J.* **2005**, *10*, 59–69. [**Google Scholar**]

381. Huang, J.K.; Ji, X.Q. The verification of the right to use farmland and farmers' long-term investment in farmland. *J. Manag. World* **2012**, *09*, 76–81+99+187–188. [**Google Scholar**]

382. Jacoby, H.G.; Li, G.; Rozalle, S. Hazards of Expropriation: Tenure Insecurity and Investment in Rural China. *Am. Econ. Rev.* **2002**, *92*, 1420–1447. [**Google Scholar**] [**CrossRef**][**Green Version**]

383. He, L.Y.; Huang, J.K. Stability of land use rights and fertilizer use: An empirical study in Guangdong Province. *China Rural. Surv.* **2001**, *5*, 42–48+81. [**Google Scholar**]

384. Chen, T.; Meng, L.J. Land Adjustment, Land Rights Stability and Long-term Investment of Farmers: An Empirical Analysis Based on Survey Data from Jiangsu Province. *Issues Agric. Econ.* **2007**, *10*, 4–11+110. [**Google Scholar**]

385. He, H.R.; Zhang, L.X.; Li, Q. Study on farmers' fertilization behavior and agricultural non-point source pollution. *J. Agrotech. Econ.* **2006**, *6*, 2–10. [**Google Scholar**]

386. Feng, G.; Wang, P.Z.; Zheng, L.S. Study on the effect of organic fertilizer on increasing resistance of spring maize in dryland. *J. Maize Sci.* **1998**, *S1*, 90–92. [**Google Scholar**]

387. Sun, Y.; Gao, Y.S.; Zhu, Z.Y.; Cui, H.C.; Ning, L.R.; Yan, X.G. Research on drought resistance and yield enhancement technology of maize in the semi-arid area of western Jilin Province. *Agric. Res. Arid. Areas* **2002**, *3*, 7–11. [**Google Scholar**]

388. Gan, Y.L.; Qi, L.; Li, H.; Liu, G.L. Effects of organic fertilizer application on the occurrence and yield of major pests and diseases in rice. *J. Zhejiang Agric. Sci.* **2014**, *12*, 1844–1846. [**Google Scholar**]

389. Hou, J.D.; Lv, J.; Yin, W.F. Study on the impact of farmers' business behavior on the rural ecological environment. *China Popul. Resour. Environ.* **2012**, *22*, 26–31. [**Google Scholar**]

390. Kassie, M.; Zikhali, P.; And, K.M.; Edwards, S. Adoption of organic farming techniques: Evidence from a semi-arid region of Ethiopia. *Environ. Dev. Discuss. Pap.-Resour. Future RFF* **2009**, *9*, 297–321. [**Google Scholar**]

391. Zemedu, B. Analysis of the impact of organic fertilizer use on smallholder farmers' income in Shashemene District, Ethiopia. *Int. J. Agric. Econ.* **2016**, *1*, 117–124. [**Google Scholar**]

392. Song, D.P.; Chen, W.; Gao, Y.Z. The usage rationality and environmental impacts of chemical nitrogen fertilizer and pesticide in the Huaihe River Basin, China. *J. Agro-Environ. Sci.* **2011**, *30*, 1144–1151. [**Google Scholar**]

393. Yu, G.; Zhou, X.B. Research on the impact of agricultural fertilizers and bio-organic fertilizers on the environment. *Agric. Econ.* **2022**, *421*, 12–14. [**Google Scholar**]

394. Huang, Y.Z.; Luo, X.F. Reduction and substitution of fertilizers: Farmer's technical strategy choice and influencing factors. *J. South China Agric. Univ. Soc. Sci. Ed.* **2020**, *19*, 77–87. [**Google Scholar**]

395. Li, H.; Lu, Q. Can product quality certification improve farmers' technological efficiency. *Chin. Rural. Econ.* **2020**, *425*, 128–144. [**Google Scholar**]

396. Li, G.C. The composition, changes and distribution of China's agricultural productivity growth in the transitional periods. *China Popul.,Resour. Environ.* **2009**, *19*, 148–152. [**Google Scholar**]

397. Yuan, X.L.; Wang, J.; Zhang, J.Y. Evaluation of urban efficiency under high-quality development based empirical research from 19 cities at sub-provincial level and above. *Urban Dev. Stud.* **2020**, *27*, 62–70. [**Google Scholar**]

398. Qin, W. In the context of the construction of the Guangdong-Hong Kong-Macao Greater Bay Area evaluation of modern logistics efficiency and analysis of influencing factors in Zhuhai. *China Econ. Trade Her.* **2021**, *992*, 66–68. [**Google Scholar**]

399. Han, Y.D.; Liu, H.W. Study on technology efficiency of logistics industry in China and influence factors: An empirical study from listed companies. *China Bus. Mark.* **2019**, *33*, 17–26. [**Google Scholar**]

400. Huang, Q.H.; Shi, P.H.; Hu, J.F. Industrial agglomeration and high- quality economic development: Examples of 107 prefecture- level cities in the Yangtze River Economic Belt. *Reform* **2020**, *311*, 87–99. [**Google Scholar**]

401. Yang, W.J.; Li, Q. Study on effect of new business entities' productive services on rice production technical efficiency: Based on household survey data of 1965 farmers in 12 provinces. *J. Huazhong Agric. Univ. Soc. Sci. Ed.* **2017**, *131*, 12–19+144. [**Google Scholar**]

402. Peng, D.Y.; Wu, X. A study on China's agricultural technology efficiency and TFP: A perspective on the changes in the structure of rural labor force. *Economist* **2013**, *9*, 68–76. [**Google Scholar**]

403. Zhang, H.; Zhang, Y.M. Does agricultural subsidy improve technical efficiency of grain production: An empirical study based on the data of 552 grain production family farms in Jiangsu Province. *J. Huazhong Agric. Univ. Soc. Sci. Ed.* **2022**, *162*, 58–67. [**Google Scholar**]

404. Wu, H.X.; Hao, H.T.; Shi, H.T.; Ge, Y. Effect of agricultural mechanization on total factor productivity of wheat and its spatial spillover effect. *J. Agrotech. Econ.* **2022**, *328*, 50–68. [**Google Scholar**]

405. Tian, Z.; Wang, R.; Shi, Y. Technical efficiency of grain production in family farm with different scales in developed areas: A survey of family farm from Songjiang, Shanghai. *Chin. J. Agric. Resour. Reg. Plan.* **2022**, *43*, 150–159. [**Google Scholar**]

406. Cui, Z., Zhang, H., Chen, X., Zhang, C., Ma, W., Huang, C., ... & Dou, Z. (2018). Pursuing sustainable productivity with millions of smallholder farmers. *Nature*, 555(7696), 363-366.

407. Erisman, J. W., Sutton, M. A., Galloway, J., Klimont, Z., & Winiwarter, W. (2008). How a century of ammonia synthesis changed the world. *Nature Geoscience*, 1(10), 636-639.

Smil, V. (2004). Enriching the Earth: Fritz Haber, Carl Bosch, and the Transformation of World Food Production. MIT Press.

408. If crop yields had remained at their levels in 1961, we would need almost three times as much farmland today (to meet food production in 2019). Crop yield gains have 'saved' 1.7 billion hectares of land. That's equal to an area the size of the USA and Brazil combined.

409. By 'football', I mean 'soccer'.

410. Mueller, N. D., Gerber, J. S., Johnston, M., Ray, D. K., Ramankutty, N., & Foley, J. A. (2012). Closing yield gaps through nutrient and water management. *Nature*, 490(7419), 254-257.

411. Lassaletta, L., Billen, G., Grizzetti, B., Anglade, J., & Garnier, J. (2014). 50 year trends in nitrogen use efficiency of world cropping systems: the relationship between yield and nitrogen input to cropland. *Environmental Research Letters*, 9(10), 105011.

412. Wuepper, D., Le Clech, S., Zilberman, D., Mueller, N., & Finger, R. (2020). Countries influence the trade-off between crop yields and nitrogen pollution. *Nature Food*, 1(11), 713-719.

413. Wuepper, D., Le Clech, S., Zilberman, D., Mueller, N., & Finger, R. (2020). Countries influence the trade-off between crop yields and nitrogen pollution. *Nature Food, 1*(11), 713-719.

414. In the few cases that natural vegetation potential did vary across borders, this was corrected for in the results.

415. Kurdi, Sikandra; Mahmoud, Mai; Abay, Kibrom A.; and Breisinger, Clemens. 2020. Too much of a good thing? Evidence that fertilizer subsidies lead to overapplication in Egypt. MENA RP Working Paper 27. Washington, DC: International Food Policy Research Institute (IFPRI). https://doi.org/10.2499/p15738coll2.133652.

416. Finger, R., Swinton, S. M., El Benni, N., & Walter, A. (2019). Precision farming at the nexus of agricultural production and the environment. Annual Review of Resource Economics, 11, 313-335.

417. Walter, A., Finger, R., Huber, R., & Buchmann, N. (2017). Opinion: Smart farming is key to developing sustainable agriculture. *Proceedings of the National Academy of Sciences*, 114(24), 6148-6150.

Prof. Dr. Hayfaa Jasim Hussein

Department of Soil Science and Water Resources, College of Agriculture, University of Basrah, Iraq

E-mail: hayfaa.hussein@uobasrah.edu.iq

Haifa.jasim@yahoo.com

altamimi.hayfaa1@gmail.com

Tell: (+964) 7703162753

❖ Prof. Dr. Hayfaa Jasim Hussein Al-Tameemi

❖ Completed primary, intermediate and secondary education in Basrah Governorate / Iraq

❖ Obtained a bachelor's degree in plant production / soil branch / College of Agriculture / Basrah University / Iraq / 1982

❖ Obtained a master's degree in Soil Science / Fertilizers and Soil Fertility / College of Agriculture / University of Basrah / Iraq / 1988

❖ Obtained a PhD in Soil Sciences / Fertilizers and Soil Fertility / College of Agriculture / Basrah University / Iraq / 1998

❖ Got the title of Professor from the University of Basrah / College of Agriculture / Soil Department on 9/25/2004

❖ Expert at the Department of Soil Sciences and Water Resources from 2017 until now

❖ Supervised a number of master's students (11) and doctoral students (13)

❖   Co-author a book (The Impact of Climate Change on Soil and Its Components) / 2023(with Prof.Dr.Mohamed Abdel-Raheem Ali Abdel-Raheem from Egypt).

❖   Published more than 75 research papers in scientific and international journals

❖   Member of several ministerial, university, college and department committees

❖   Participated in many conferences, seminars and workshops inside and outside Iraq

❖   Member of the environmental team at the University of Basrah/Iraq

❖   Held the position of chairperson of the academic promotions committee in the college for seven years (2006-2013)

❖   Member of the Editorial Board of the Iraqi Journal of Soil Sciences

❖   Held the position of Head of the Department of Soil Sciences and Water Resources from 7/15/2020 until now.

❖   Member of the Iraqi Academics Syndicate

Prof. Dr. Mohamed Abdel-Raheem Ali Abdel-Raheem

Professor of Entomology (Biological Control)

Pests & Plant Protection Department

Agricultural and Biological Research Institute

National Research Centre.

33^{d} ElBohouth St., Dokki, Cairo, Egypt.

Email: abdelraheem_nrc@hotmail.com,

abdelraheem_nrc@yahoo.com

Mobile: (+2) 01155527583 - (+2) 0100958079

Prof. Dr. Mohamed Abdel-Raheem Ali Abdel-Raheem, Prof. of Entomology (Biological Control), Pests & Plant Protection Department, Agricultural and Biological Research Institute, National Research Centre, Cairo, Egypt. Published publications (240) papers (73), Books and Book chapters (167), (Scopus) h-index (9), Citations   (218), (Google Scholar) h-index (15) , Citations (707), (Web of Science) h-index (3), Citations (23), Research Gate, h-index (11), Citations   (471), Reviewers in International Journals (108), Reviewed articles in Biological Control (380), Editor in Chief in Journal (5), Editorial board in Journals (29), Associated Editor in Chief in Journal (5), conference Attended (29), workshop Attended (289), Symposium Attended (179), Forum Attended (4), Prize (4), Attended Others (30), Projects As PI and member (16),   Training courses For Agric. Engineering (18), Training courses For Students of University (8), Attending Trainees courses (15), TV & Radio (16), Supervisor on Ph.D. thesis (1), Committee Ph.D. thesis (2), member in scientific Foundation (13).

National Research Centre. 33rd ElBohouth St., Dokki, Cairo, Egypt. Email Address: abdelraheem_nrc@hotmail.com,   abdelraheem_nrc@yahoo.com, ma.abdel-raheem@nrc.sci.eg, Mobile Phone: (+2) 01155527583 – (+2) 01009580797

Printed by Books on Demand GmbH, Norderstedt / Germany